Advances in Science, Technology & Innovation

IEREK Interdisciplinary Series for Sustainable Development

Editorial Board

Anna Laura Pisello, Department of Engineering, University of Perugia, Italy

Dean Hawkes, University of Cambridge, Cambridge, UK

Hocine Bougdah, University for the Creative Arts, Farnham, UK

Federica Rosso, Sapienza University of Rome, Rome, Italy

Hassan Abdalla, University of East London, London, UK

Sofia-Natalia Boemi, Aristotle University of Thessaloniki, Greece

Nabil Mohareb, Faculty of Architecture - Design and Built Environment, Beirut Arab University, Beirut, Lebanon

Saleh Mesbah Elkaffas, Arab Academy for Science, Technology, Egypt

Emmanuel Bozonnet, University of la Rochelle, La Rochelle, France

Gloria Pignatta, University of Perugia, Italy

Yasser Mahgoub, Qatar University, Qatar

Luciano De Bonis, University of Molise, Italy

Stella Kostopoulou, Regional and Tourism Development, University of Thessaloniki, Thessaloniki, Greece

Biswajeet Pradhan, Faculty of Engineering and IT, University of Technology Sydney, Sydney, Australia

Md. Abdul Mannan, Universiti Malaysia Sarawak, Malaysia

Chaham Alalouch, Sultan Qaboos University, Muscat, Oman

Iman O. Gawad, Helwan University, Egypt

Anand Nayyar, Graduate School, Duy Tan University, Da Nang, Vietnam

Series Editor

Mourad Amer, International Experts for Research Enrichment and Knowledge Exchange (IEREK), Cairo, Egypt

Advances in Science, Technology & Innovation (ASTI) is a series of peer-reviewed books based on important emerging research that redefines the current disciplinary boundaries in science, technology and innovation (STI) in order to develop integrated concepts for sustainable development. It not only discusses the progress made towards securing more resources, allocating smarter solutions, and rebalancing the relationship between nature and people, but also provides in-depth insights from comprehensive research that addresses the **17 sustainable development goals (SDGs)** as set out by the UN for 2030.

The series draws on the best research papers from various IEREK and other international conferences to promote the creation and development of viable solutions for a **sustainable future and a positive societal** transformation with the help of integrated and innovative science-based approaches. Including interdisciplinary contributions, it presents innovative approaches and highlights how they can best support both economic and sustainable development, through better use of data, more effective institutions, and global, local and individual action, for the welfare of all societies.

The series particularly features conceptual and empirical contributions from various interrelated fields of science, technology and innovation, with an emphasis on digital transformation, that focus on providing practical solutions to **ensure food, water and energy security to achieve the SDGs**. It also presents new case studies offering concrete examples of how to resolve sustainable urbanization and environmental issues in different regions of the world.

The series is intended for professionals in research and teaching, consultancies and industry, and government and international organizations. Published in collaboration with IEREK, the Springer ASTI series will acquaint readers with essential new studies in STI for sustainable development.

ASTI series has now been accepted for Scopus (September 2020). All content published in this series will start appearing on the Scopus site in early 2021.

More information about this series at https://link.springer.com/bookseries/15883

Fatimazahra Barramou •
El Hassan El Brirchi •
Khalifa Mansouri • Youness Dehbi
Editors

Geospatial Intelligence

Applications and Future Trends

Editors
Fatimazahra Barramou
Hassania School of Public Works
Casablanca, Morocco

El Hassan El Brirchi
Hassania School of Public Works
Casablanca, Morocco

Khalifa Mansouri
Higher Normal School of Technical
Education of Mohammedia
Mohammedia, Morocco

Youness Dehbi
Institute of Geodesy and Geoinformation
University of Bonn
Bonn, Germany

ISSN 2522-8714   ISSN 2522-8722   (electronic)
Advances in Science, Technology & Innovation
IEREK Interdisciplinary Series for Sustainable Development
ISBN 978-3-030-80460-2   ISBN 978-3-030-80458-9   (eBook)
https://doi.org/10.1007/978-3-030-80458-9

© The Editor(s) (if applicable) and The Author(s), under exclusive license to Springer Nature Switzerland AG 2022
This work is subject to copyright. All rights are solely and exclusively licensed by the Publisher, whether the whole or
part of the material is concerned, specifically the rights of translation, reprinting, reuse of illustrations, recitation,
broadcasting, reproduction on microfilms or in any other physical way, and transmission or information storage and
retrieval, electronic adaptation, computer software, or by similar or dissimilar methodology now known or hereafter
developed.
The use of general descriptive names, registered names, trademarks, service marks, etc. in this publication does not
imply, even in the absence of a specific statement, that such names are exempt from the relevant protective laws and
regulations and therefore free for general use.
The publisher, the authors and the editors are safe to assume that the advice and information in this book are believed
to be true and accurate at the date of publication. Neither the publisher nor the authors or the editors give a warranty,
expressed or implied, with respect to the material contained herein or for any errors or omissions that may have been
made. The publisher remains neutral with regard to jurisdictional claims in published maps and institutional
affiliations.

This Springer imprint is published by the registered company Springer Nature Switzerland AG
The registered company address is: Gewerbestrasse 11, 6330 Cham, Switzerland

Preface

Currently, Geospatial Data are democratized leading to a huge of data available to everyone. Handling of such large volume of data is, however, a complex and laborious problem. Processing large volume of data requires vast computing power and specific expertise. That is why we need to use artificial intelligence that structure data, optimize productivity, and automate repetitive tasks. The main objective of this book entitled *Geospatial Intelligence: Applications and Future Trends* is to explore cutting-edge methods combining geospatial technologies and artificial intelligence related to several diversified fields such as smart farming, urban planning, transportation, and 3D city models. It introduces techniques which range from machine and deep learning to remote sensing for geospatial data analysis. The work in this book is addressed to the scientific community interested in the coupling between Geospatial Technologies and Artificial Intelligence. It will offer practitioners and researchers from academia, industry and government information, experiences, and research results about all aspects of specialized and interdisciplinary fields on Geospatial Intelligence. This book will focus on the recent trends and challenges for employing artificial intelligence for remote sensing and spatial data analytics. To achieve the objectives, the book consists of two main parts that includes 13 chapters contributed by promising authors.

The first part deals with the use of artificial intelligence techniques to improve spatial data analytics. It includes 7 chapters: In chapter 1, "Towards a Multi-agents Model for Automatic Big Data Processing to Support Urban Planning", Sassite et al. propose a multi-agent model for automating big data processing based on a smart data approach. In chapter 2, "Geospatial Forecasting and Social Media Exploration Based on Sentiment Analysis: Application to Flood Forecasting", Abas et al. propose a geospatial forecasting approach to analyze the textual content of social media using geospatial components and Natural Language Processing (NLP) tools. In chapter 3, "Deep Convolution Neural Network for Automated Method of Road Extraction on Aerial Imagery", Norelyaqine et al. present the use, deployment, and validation of deep learning strategies, in particular, the U-net architecture based on deep convolutional neural networks for extracting roads from remote sensing images. In chapter 4, "Enhancing the Management of Traffic Sequence Following Departure Trajectories", Bikir et al. establish an algorithm that optimizes the departure sequence taking into account the aircraft categories, the spent time on the taxiway, the runway, and the climb in the departure trajectory. In chapter 5, "A Multiagent and Machine Learning Based Denial of Service Intrusion Detection System for Drone Networks", Ouizzane et al. propose a new model based on Muli-Agent System and on machine learning techniques to detect DoS cyber-attacks targeting drones networks. In chapter 6, "Toward a Deep Learning Approach for Automatic Semantic Segmentation of 3D Lidar Point Clouds in Urban Areas", Ballouch et al. propose a new approach based on the combination of Lidar data and other sources in conjunction with a deep dearning technique to automatically extract semantic information from airborne Lidar point clouds by enhancing both accuracy and semantic precision compared to the existing methods. In chapter 7, "Artificial and Geospatial Intelligence Driven Digital Twins' Architecture Development Against the Worldwide Twin Crisis Caused by COVID-19", Mezzour et al. propose a generic solution combining advanced simulation and multidimensional modeling through DT, geostatistical exploration through location intelligence and new business models by BI for strengthening the resilience of critical value chains.

The second part deals with the use of artificial intelligence with remote sensing in several fields. It includes six chapters: in chapter 8, "Opportunities for Artificial Intelligence in Precision Agriculture Using Satellite Remote Sensing", Dakir et al. present recent techniques, algorithms, and methodologies using artificial intelligence in precision agriculture using satellite remote sensing, and concern recent studies that were conducted in latest years 2019–2020. In chapter 9, "Monitoring Land Productivity Trends in Souss-Massa Region Using Landsat Time Series Data to Support SDG Target 15.3", Saadani et al. conducted a study for measuring the proportion of degraded land in the Souss–Massa region by the use of Google Earth Engine. In chapter 10, "Subimages-Based Approach for Landslide Susceptibility Mapping Using Convolutional Neural Network", Alami Machichi et al. developed a convolutional neural networks (CNN) model capable of producing a susceptibility map using seven explanatory variables: lithology, slope, drainage density, fault density, elevation, roughness, and aspect. A susceptibility index map was generated in the Aknoul region in the Rif to illustrate the CNN results. In chapter 11, "Lithological Mapping for a Semi-arid Area Using GEOBIA and PBIA Machine Learning Approaches with Sentinel-2 Imagery: Case Study of Skhour Rehamna, Morocco", Serbouti et al. conduct a study to evaluate different approaches including pixel and object-based image analysis, in order to select the most accurate approach for mapping lithological units in semi-arid areas, where Skhour Rehamna was chosen as a case study region. In chapter 12, "Optimization of Object-Based Image Analysis with Genetic Programming to Generate Explicit Knowledge from WorldView-2 Data for Urban Mapping", Azmi et al. conduct a study that aims to examine how data mining techniques, in the context of rule induction algorithms based on GP method, can help to discover knowledge and extract classification rules automatically to well illustrate this knowledge. In chapter 13, "Machine Learning and Remote Sensing in Mapping and Estimating Rosemary Cover Biomass", Chafik et al. present an efficient method to estimate the bio-mass of rosemary cover based on satellite imagery data and machine learning technics.

Casablanca, Morocco   Prof. Fatimazahra Barramou
Casablanca, Morocco   Prof. El Hassan El Brirchi
Mohammedia, Morocco   Prof. Khalifa Mansouri
Bonn, Germany   Dr.-Ing. Youness Dehbi

Contents

Contributors

Sara Abas Architecture, System and Networks Team (ASYR) - Laboratory of Systems Engineering (LaGeS), Hassania School of Public Works EHTP, Casablanca, Morocco

Bikir Abdelmounaime Laboratory Signals, Distributed Systems and Artificial Intelligence ENSET, University Hassan II, Mohammedia, Morocco

Norelyaqine Abderrahim Laboratoire de géophysique appliquée, de géotechnique, de géologie de l'ingénieur et de l'environnement (L3GIE), Mohammed V University – Mohammadia School of Engineering, Rabat, Morocco

Malika Addou ASYR RT, LaGes Laboratory, Hassania School of Public Works, Casablanca, Morocco

Omar Bachir Alami Geomatics Science Research team (SGEO), LaGeS Laboratory, Hassania School of Public Works, Casablanca, Morocco

Aouatif Amine Department of Informatics, Logistics, and Mathematics, National High School of Applied Sciences, University Ibn Tofail, Kenitra, Morocco

Z. Ballouch College of Geomatic Sciences and Surveying Engineering, IAV Hassan II, Rabat, Morocco

Fatimazahra Barramou Geomatics Science Research Team (SGEO), LaGeS Laboratory, Hassania School of Public Works, Casablanca, Morocco

Mohamed Berrada Department of Mathematics and Informatics, National High School of Arts and Crafts, University of Moulay Ismail, Meknes, Morocco

Hassan Chafik Department of Mathematics and Informatics, National High School of Arts and Crafts, University of Moulay Ismail, Meknes, Morocco

Asmae Dakir Geomatics Science Research team (SGEO), LaGeS Laboratory, Hassania School of Public Works, Casablanca, Morocco

M. Ettarid College of Geomatic Sciences and Surveying Engineering, IAV Hassan II, Rabat, Morocco

Mezzour Ghita National and High School of Electricity and Mechanic (ENSEM) HASSAN II University, Casablanca, Morocco;
Research Foundation for Development and Innovation in Science and Engineering, Casablanca, Morocco;
Innovation Lab for Operations (ILO), Mohammed VI Polytechnic University (UM6P), Morocco, Ben Guerir, Morocco

Peter L. Guth US Naval Academy, Annapolis, MD, USA

Griguer Hafid Innovation Lab for Operations (ILO), Mohammed VI Polytechnic University (UM6P), Morocco, Ben Guerir, Morocco

R. Hajji College of Geomatic Sciences and Surveying Engineering, IAV Hassan II, Rabat, Morocco

Mustapha Hakdaoui Department of Geology, Laboratory of Applied Geology, Geomatic and Environment, Faculty of Sciences Ben M'Sik, Hassan II University of Casablanca, Casablanca, Morocco

El Hassan El Brirchi LaGeS Laboratory, Geomatics Science Research Team'SGEO', Hassania School of Public Works, Casablanca, Morocco

Amar Hicham Mohammed VI Polytechnic University, Mining Environment & Circular Economy, Ben Guerir, Morocco

Medromi Hicham National and High School of Electricity and Mechanic (ENSEM) HASSAN II University, Casablanca, Morocco;
Research Foundation for Development and Innovation in Science and Engineering, Casablanca, Morocco

Khalifa Mansouri Laboratory Signals, Distributed Systems and Artificial Intelligence ENSET, University Hassan II, Mohammedia, Morocco

Said Lahssini National Forestry School of Engineers, Sale, Morocco

Anass Legdou Department of Informatics, Logistics, and Mathematics, National High School of Applied Sciences, University Ibn Tofail, Kenitra, Morocco

Mouad Alami Machichi Moroccan Foundation for Advanced Science, Innovation & Research, Rabat, Morocco

Saadani Moussa LaGeS Laboratory, Geomatics Science Research Team'SGEO', Hassania School of Public Works, Casablanca, Morocco

Idrissi Otmane Laboratory Signals, Distributed Systems and Artificial Intelligence ENSET, University Hassan II, Mohammedia, Morocco

Said Ouiazzane ASYR RT, LaGeS Laboratory, Hassania School of Public Works, Casablanca, Morocco

Mohammed Raji Department of Geology, Laboratory of Applied Geology, Geomatic and Environment, Faculty of Sciences Ben M'Sik, Hassan II University of Casablanca, Casablanca, Morocco

Azmi Rida Center of Urban Systems, Mohamed VI Polytechnic University (UM6P), Ben Guerir, Morocco

Abderrahim Saadane Department of Geology, Faculty of Sciences of Rabat, University Mohammed V, Rabat, Morocco

Fouad Sassite ASYR RT, LaGeS Laboratory, Hassania School of Public Works, Casablanca, Morocco

Imane Serbouti Department of Geology, Laboratory of Applied Geology, Geomatic and Environment, Faculty of Sciences Ben M'Sik, Hassan II University of Casablanca, Casablanca, Morocco

Benhadou Siham National and High School of Electricity and Mechanic (ENSEM) HASSAN II University, Casablanca, Morocco;
Research Foundation for Development and Innovation in Science and Engineering, Casablanca, Morocco

Spatial Data and Artificiel Intelligence

Towards a Multi-agents Model for Automatic Big Data Processing to Support Urban Planning

Fouad Sassite, Malika Addou, and Fatimazahra Barramou

Abstract

The objective of this paper was to propose a multi-agents model for automating big data processing, to improve the process of decision-making and urban planning. The huge amounts of collected data from different domains, such as urban management and remote sensing, are characterized as big data with a spatial component. Smart data is the approach to deal with big data characteristics and challenges by focusing on the Value aspect. The focus on smart data on the relevant data and the mechanism of automation and collaboration of the agents, will contribute to the efficient automating for big data analytics and processing. The proposed approach is based on a collection of agents, and adopt the concept of smart data, this paradigm focus on the aspect value from the big data and help to retrieve the useful information from the large volumes of data by intelligent processing. The proposed model describes the functionalities of the agents. The agents receive data in real time, perform the operations of storing data, pre-processing, streaming processing and batch processing and finally transfer the results of analysis to the services and applications. Machine learning techniques can be used to enhance the aspect of cognition of the agents; through a case study, we used supervised learning methods to build a classification model to support the process of urban planning by predicting the type of habitat adequate for the population based on their demographic and socio-economic characteristics.

Keywords

Smart data • Big data • Multi-agent system • Automation • Machine learning • Urban planning • Big data analytics • Decision-making

1 Introduction

With the technological advances we are witnessing today, which are transforming our daily lives and changing the way we interact with the world around us, these advances have particularly evolved in the field of data science, through the rapid evolution of connected objects, computing power, data storage and processing.

The diversity of data sources (social networks, sensors, connected objects…) provide a large volume of data characterized by the diversity in types and formats with possible heterogeneous data.

The huge amounts of collected data from different domains, such as urban management and remote sensing, are characterized as big data with a spatial component [1]; to deal with the storage, management and processing of this data, we need reliable methods. Smart data is the approach to deal with big data characteristics and challenges by focusing on the Value aspect [2] by extracting useful information from the large volumes of big data and to minimize the latency to support real-time data processing. Another aspect of smart data is the focus on supporting decision-making and automatization.

The Big Data features increasingly generated at high rates come from different sources or systems within intelligent cities, such as traffic management data based on wireless sensor networks [3], and in different domains such as the domain of smart weather prediction [4].

The need of effective urban planning is highlighted to meet the challenge of the continuous population growth in urban areas, and it is necessary to develop a strategic and effective data system that can consider these populations in terms of planning.

F. Sassite (✉) · M. Addou
ASYR RT, LaGes Laboratory, Hassania School of Public Works, Casablanca, Morocco

F. Barramou
Geomatics Science Research Team (SGEO), LaGeS Laboratory, Hassania School of Public Works, Casablanca, Morocco

© The Author(s), under exclusive license to Springer Nature Switzerland AG 2022
F. Barramou et al. (eds.), *Geospatial Intelligence*, Advances in Science, Technology & Innovation,
https://doi.org/10.1007/978-3-030-80458-9_1

By adopting an approach that includes the cleaning, processing and analysis of data for decision-making purposes to face the different Data available for development include population index, national boundary data and satellite image information [5].

The smart data is a topic of interest to researchers since advances in big data processing; in this paper, we are interested on a smart data filed, which is automation of big data processing.

The purpose of this paper is to present a multi-agent system model to automate big data analytics and decision-making, such a need is expressed in several areas, particularly in the area of urban management and planning. The use of some techniques of machine learning used in this proposition will improve the data processing, the transition from Big data to smart data and process of decision-making.

The document is organized as follows: in Sect. 2, we will discuss some concepts and paradigms related to the problems studied. Section 3 highlights some related works. Section 4 presents the proposed model. Section 5 studies the experimentation conducted in this work, and in the final section, we will conclude and present some perspectives.

2 Background of the Study

In this section, we will highlight some concepts and paradigms related to the problem under study. First, we will begin by identifying the characteristics of geospatial big data, the added value of big data by adopting the smart data approach, then we will present algorithms of machine learning used to build the model in the case study, finally we will talk about the multi-agent paradigm.

The objective is to build a solution capable to automate big data analytics, based on the smart data approach.

2.1 Big Data

The massive production of data that we witness today is a result of the rapid evolution of storage media, connected objects, sensors, online services, etc. These quantities of data generated at high rates are mainly characterized by the high volume and complexity of data sets. This characteristic is a result of the diversity of data sources; consequently, there is a need to analyze the data flows with efficient techniques to better benefit from the generated data [2].

2.2 Geospatial Big Data

Generally, the term of big data try to describe the data characterized with complexity in different terms, in either storage, management or analysis [6].

Systems dealing with Geospatial big data manage data in different type structures, at the same time, lead a real-time analysis to help in the process of obtaining valuable data that will be useful for further decision-making.

The geospatial Big data is characterized with 7v related to the already available description of big data in addition to other characteristics of the spatial aspect of this type of data.

The volume characteristic is the most discussed when the subject of big data is highlighted, this characteristic tries to describe other notions like the size of data, the storage organization and strategies. Velocity describes the speed of data generation, and the real-time data management and processing. Variety includes the aspect data type, in different form and structuration. Veracity stands for data integrity, the gathered data should correspond to the result of the data processing. Versatility characteristic is interested in development and support of user-friendly interfaces to show and expose the results of processing data (Fig. 1).

2.3 Smart Data

The aim of smart data is to provide useful information from big data described by large quantities of data by concentrating efforts on the value aspect and the veracity of data, by filtering or transforming data and to reveal the valuable aspect which can be used in different applications by governments and businesses like improve decision-making and planning tasks [8].

The data used in order to be smart should have three main characteristics:

- Actionable: The studied data should lead to take actionable actions, based on this value-driven approach.
- Accurate: Data accuracy is an important aspect of data quality and it means that the data value stored should be the correct value; this representation should be consistent and clear to avoid dealing with ambiguity.
- Agile: data availability should be guaranteed in real time, and the evolution of data can affect the organizations and lead them to change in a flexible way.

In the smart data projects, we are particularly interested in combining advancement in the next four fields in order to build reliable solutions [9]:

- Reliable Infrastructure.
- Data Organization and Management.
- Analysis and Prediction.
- Decision Support and Automatization.

2.4 Machine Learning

Machine learning is a type of Artificial intelligence and a sub-domain of computer science that allows machines to learn without explicit programming. Machine learning has developed essentially from computer learning theory and pattern

Fig. 1 7Vs: characteristics of geospatial big data. Redrawn based on [7]

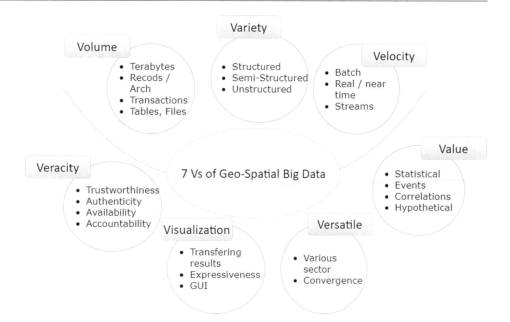

recognition. Generally, there are three major categories of machine learning: unsupervised, supervised and reinforcement [10].

In the supervised learning, the entire training consists of samples of data input vectors and their corresponding appropriate target vectors, commonly known as labels. The goal of supervised learning is to learn to predict the corresponding output vector for a given input vector, this type of algorithms can be used to solve problems that necessitate classification tasks. In the case of unsupervised learning, no labels are required for the training set. Machine learning by reinforcement addresses to the problem of learning the appropriate action or sequence of actions to be taken for a given situation in order to maximize the score [11].

Reaching meaningful results by processing big data goes through different stages from data acquisition to consuming services and reporting. Machine learning can be a very useful technique in order to get insights and discover hidden patterns in data. After the acquisition of data, a series of tasks is performed like pre-processing data and applying filters and cleaning data, and split the data into training set and testing set, the result can be very helpful in the process of decision-making.

2.4.1 Supervised Machine Learning

The Supervised Machine Learning process consists of different consecutive steps aimed at building a model that will be used as a classifier. The first step is closely related to the trade studied where the problem to be solved using Machine Learning techniques must be identified, then the identification of the data needed to build the model and subsequently solve the problem studied, then we have to go through the pre-processing stage. This stage is very important in the

learning process because it allows us to prepare the data according to the necessary requirements; in this part, we can apply the tasks, and to start the learning part, we have to divide the data into two datasets: one part for learning and the other for testing the model.

The next step is the choice of the algorithm: this choice depends on the type of problem, the learning objective and the nature of the data.

After the training phase of the model, the model will be validated in the evaluation stage, thanks to the Test Data Set. If the model is not reliable, the parameters used in the training phase must be changed or adjusted, then after having a reliable model, it can be saved and used in any needed prediction (Fig. 2).

2.4.2 Semi-supervised Machine Learning

Semi-supervised learning is mainly based on the two types of learning, the supervised learning that uses data with labels and the unsupervised learning that uses unlabeled data. It is a combination between the two to build a solution that takes advantage of the benefits of each technique. This type of learning is often used in situations where we process large volumes of data without labeling them all [13].

The data-labeling phase can be a very time- and resource-consuming task, especially if an expert intervention is required and the data to be labeled is large. We mainly talk about:

- Transductive learning: whose objective is to make predictions about observations in order to minimize error, it can also provide the class for already labeled data.
- Inductive learning: this type of learning is able to provide for unlabeled data, the classifier and labels; it is based on other learning methods such as self-learning, S3VM and T-SVM.

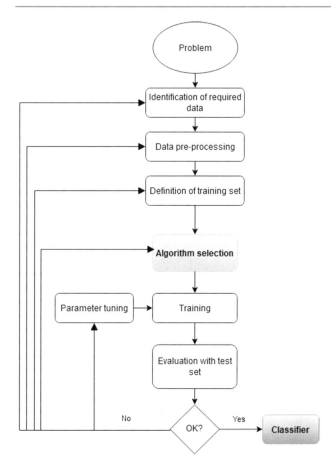

Fig. 2 Supervised machine learning workflow. Redrawn based on [12]

2.4.3 Applications of Machine Learning

The applications of machine learning algorithms are very numerous and they are used in several fields and disciplines. We find applications of these techniques for gesture recognition, improving the experience of human–computer interaction, computer vision, prediction, semantic analysis, recommendation, intrusion detection, object recognition, natural language processing, social media… [14]. This technique also finds its application in the geospatial field in different domains like urban planning, self-driving techniques for vehicles and disaster management [15].

The applications of this technique is widely used in different areas by giving the ability of learning to the machines, and by taking advantages from the powerful capacity of computing and storing data that we are witnessing today.

In our study, we had to use some algorithms for supervised learning with the aim to get a classifier. We studied various algorithms in this field like: decision tree, naïve Bayes, Random Forest, Adaptive Boosting, support Vector Machine (SVM) and the multi-layer perceptron (MLP) for classification.

We will present here the three main algorithms we used in the study in the previous section, and then we will present the metrics, used to choose the most accurate model:

- Naïve Bayes:

The classifiers based on Bayes' naive methods are a group of supervised learning algorithms, which are based on the use of the Bayes' theorem, with the "naïve" hypothesis of a conditional independence of each pair of characteristics given the value of the class variable. The Bayes theorem establishes the following relationship, given the class variable and dependent characteristic vector [16].

The Bayes rule:

$$P(y \mid x_1, \ldots, x_n) = \frac{P(y)P(x_1, \ldots, x_n \mid y)}{P(x_1, \ldots, x_n)} \qquad (1)$$

- Random Forest:

The principle of the Random Forest classifier is a combination of the concepts of random subspaces and bagging. The random forest algorithm performs a division of the training data into a specific numbers of subsets of data used to perform learning on multiple decision trees, the used subsets used to train the model are slightly different, then the forest based on the method of vote choose the classification with the most votes [17] (Fig. 3).

- Multi-Layer Perceptron (MLP):

The Multi-layer perceptron MLP is a type of artificial neural network organized as a set of several layers, the information only flows from the input layer to the output layer. It is therefore a feedforward network, each layer is constituted of a variable number of neurons, the neurons of the last layer being the outputs of the global system. The Multi-layer Perceptron can be used either for scenario of classification or regression, and to optimize models. Based on the MLP, we can use various algorithms such as: Adam, SGD and L-BFGS. Some of the strengths of this model are:

- The possibility to learn and build the models in real-time.
- The possibility to use it for non-linear models [19].

2.4.4 Metrics

To evaluate the performance of the classifiers and as a systematic comparison approach, there are typically four used measures of precision in a classification problem [20]. We will use in this study the following metrics: Accuracy, Precision, Recall and F1-Score, these metrics are calculated on the basis of the following parameters:

TP: True Positives,
FP: False Positives,

Fig. 3 Random forest algorithm. Redrawn based on [18]

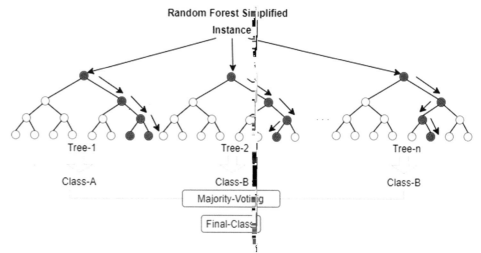

TN: True Negatives,
FN: False Negatives.

Actual class	Predicted class	
	False	True
False	TN	FP
True	FN	TP

- Accuracy is the most intuitively measured measure of performance. It represents the ratio of the correctly predicted observation to the set of observations:

$$Accuracy \ = \ TP + TN \ / \ TP + FP + FN + TN \quad (2)$$

- Precision is the ratio of well-predicted positive observations to the total predicted observations:

$$Precision \ = \ TP/TP + FP \quad (3)$$

- Recall is the ratio of well-predicted positive observations to all observations in the actual class:

$$Recall \ = \ TP/TP + FN \quad (4)$$

- F1-Score is the weighted average of accuracy and recall. This score therefore takes into consideration both false positives and false negatives.

$$F1 \ Score \ = 2$$
$$* (Recall \ * \ Precision) \ / \ (Recall \ + \ Precision) \quad (5)$$

2.5 Multi-agent System

According to [24], agent is an entity that is located in an environment and senses the several parameters that are used to reach a decision according to the entity's objective. The entity completes the necessary environmental measures based on the taken decision, and the type of this entity can be a physical like robots, hardware components or virtual one like software agents.

An agent is in interaction with its environment, which is where it is located; the environment has a major role in the complexity of the system based on multi-agent paradigm.

Some of the interesting functionalities of the agents that show their capabilities to complete some tasks in a collaborative way,

- Sociability: the agents are capable of sharing their knowledge and send requests to get informations from other agents, in order to improve reaching their objectives.
- Autonomy: every single agent can execute its own decision-making process and try to reach its own goals and take the appropriate actions.
- Proactivity: the agents can store their own historical data, some data about other agents and their sensed data to take action in their environments; based on this information, agents can take different actions based on their located environment.

In this study, we are particularly interested in the ability of collaboration of the agents. Every agent maximizes its capability to reach its own objectives, so the use of multiple agents to solve a problem can be very useful to resolve complex problems. This paradigm is known as the multi-agent system.

The aim of the Multi-agent system is to solve a complex situation by the capability of collaboration of the agents, and by dividing that complex problem to small tasks, assigned to the different agents of the system.

The main characteristics that a multi-agent system offers to solve complex tasks are:

- The efficiency,
- The low cost,
- The flexibility,
- The reliability.

The efficiency we look for in the use of multi-agent system can be obtained by taking advantage of the autonomy of the agents. This autonomy, combined with an appropriate agent behavior, can produce systems that are capable of adapting, organizing, reacting to changes, … without any external guidance from an external authority [22].

3 Related Works

In this section, we will highlight some concepts and paradigms related to the problem under study. With the technological advances that we are witnessing today, the large generation of data, the increased use of tools and technology around the concept of big data, we are in the necessity to realize tasks of data processing and analysis. The need to perform real-time analysis for decision-making and to automate these analysis tasks in several domains including management and urban planning, in order to create autonomous systems capable of performing data management and processing tasks, and real-time decision-making.

In the literature, several works have been done to study data analysis and data processing flows. First, we will talk in this part about the Big Data processing and analysis processes, and then about some methods to automate this processing, then to justify our choice of the multi-agent systems adopted for the proposed approach.

3.1 Big Data Analytics Processes

The purpose of extracting valuable information from the acquired data and reveal the true potential of big amounts of data by guiding the process of decision-making through the use of different techniques of data mining and by following efficient processes of big data analytics. The process can be detailed principally in two main sub-processes containing five steps;

The first sub-process deals with the Data management and regroups the practices of collecting and recording data into the big data infrastructures, extracting, cleaning and annotating the collected data, and the last step consists of

carrying out the operations of data integration, aggregation and representation [23].

The Description of the necessary steps for modeling and implementing solutions to extract the value from Big Data is resumed mainly in the next four steps [24]:

- Gather data: includes the approaches of collecting data from different sources; in this step, the metadata can also be collected for further use.
- Load data: describe the process of loading data sets into data clusters for operations of data storage, data preparation and data association with their metadata.
- Transform data: describe the process of changing and transforming the data according to the business requirements or the data processing needs; this process can be performed, accordingly, by two approaches: through Batch data transformation or through the interactive mode by automating the data profiling or using some interactive interfaces.
- Extract data: this step describes the operations to be performed on the resulting data sets; it can be used for further data analytics, Databases integration, raw extracts or for operational reporting (Fig. 4).

3.2 Processes Automation Through Automatic Service Composition ASC

The efficient use of data analytics inside organizations can be a major factor contributing to their success and fast growth. The use of some advanced techniques in data analytics represents a real opportunity to gain helpful information, and to provide help for decision-makers [25].

The approach of Automatic Service Composition ASC aims to provide an improved way to deal with the process of decision-making by automating it, based on the composition of services and the generation of a framework containing the chosen services from the planning stage to execution stage.

The use of an approach Service-Oriented Computing offers the possibility to implement functionalities as a distributed system which is composed of various blocks, the services are an abstraction describing the functioning of a part of the system, and they can be consumed separately [26].

The main steps of the approach based on automatic composition system are [25]:

- Planning Stage: this step is dedicated to recognize functional and non-functional characteristics, and schedules a workflow whose component workstations are labeled with the properties.

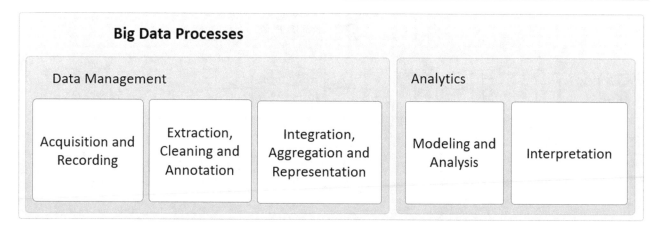

Fig. 4 Extracting insights from big data. Redrawn based on [23]

- Discovery stage: this phase tries to discover for every single task in the workflow the related services that can represent a candidate in the final workflow of processing.
- Selection stage: this phase selects the optimal service for each single task.
- Execution stage: In this phase, the previous selected services participate in the creation of the final optimized workflow using the right operators between services (Fig. 5).

3.3 Discussion

Previously, we presented some approaches to deal with data processing. Some of them describe the whole process from gathering the data from heterogeneous sources to the interpretation phase; but they do not present a mechanism to automate the whole process or to give the autonomy aspect to the system, to deal with every situation and take some decisions based on its own knowledge. In the second part, we presented an approach to automate the process based on the concept of services.

The service composition approach needs to identify in advance the whole behavior of each service and to describe in a static way all the functional behaviors. This approach can be improved by using workflow generation in a dynamic way, based on semantics where the service composition can be chosen depending on the processing.

The service composition model may also have, in an explicitly manner, the goals of the composition, to allow to the interpreter which will execute the solution, to find alternative solutions if one or more referenced services are not existing or not available.

In the literature review, we presented the multi-agent paradigm which is a powerful concept widely cited in the literature and recommended for the resolution of complex

problems in several areas such as the study of data analysis [27], the domain of robotics [28], the field of security and intrusion detection [29].

Based on the paradigm of multi-agent system, we can build solutions, to deal with the automation of data processing and overcoming the challenges in the other solutions. We can also benefit from advantages of the other solution, thanks to the capabilities of multi-agent systems to combine other technologies such as services and semantic services, the autonomous aspect of the agents, their dynamic character and their ability to learn.

4 Proposed Approach

The main idea of this paper is to present a model based on the Multi-agent and Smart Data approach to automate big data processing. This approach finds its applications in several fields, especially in urban management and planning, where we need to deal with geospatial big data management and processing in different types.

Such functionality can help in improving and automating the decision-making process, and real-time data management and processing, this work comes to complete our previous work [30].

The operating principle of the architecture proposed in the previous work is a combination between a multi-layer architecture and a multi-agent system, this architecture is mainly composed of three layers:

- The data acquisition layer: it supports the collection of data from different sources respecting the principle of data heterogeneity; in order to store these data in a cluster composed of several nodes. The system supports the management of spatial data, and handle the generated data at different rates either in batch or streaming such as the continuous collection of GPS data.

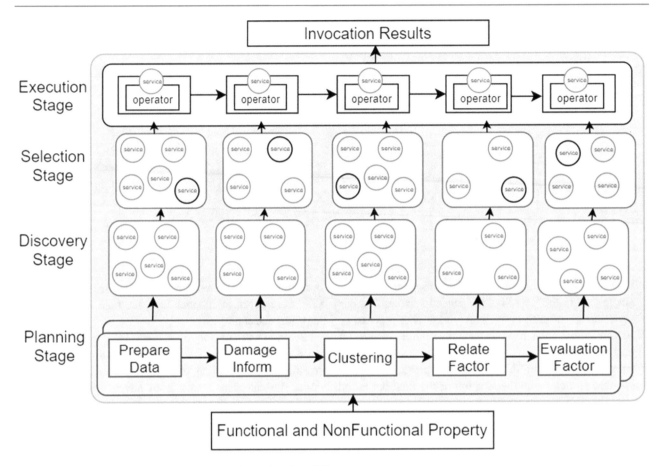

Fig. 5 ASC: Automatic Service Composition. Redrawn Based on [25]

- The data management and processing layer: This layer mainly includes the modules dedicated to data processing, namely, the components for pre-processing operations, processing queue management, data storage, batch data processing and real-time data processing.
- The data services layer: the main role of this layer is to provide processing results, dashboard tracking and reporting and also to provide interfaces to services and applications.

4.1 Overview of the Functionalities of the Proposed Multi-agent System

In this section, we will present the main functionalities of the proposed system. Which is based on a multi-agent system trying to automate the data processing and the decision management, in a big data context, this proposition adopts the smart data approach trying to deal with the complexity, the volume of big data huge datasets by focusing on the value aspect. The smart data approach can reduce the time of processing and data latency, this advantage is helpful to

improve the response of the system in processing data in real time.

The proposed Multi-agent system through the collaboration of the agents can automate the processing and management of the data on several levels: On the data acquisition component, data management and processing and on the service management and communication component.

The operation of the proposed multi-agent system (Fig. 6).

The workflow below explains the functioning of the proposed multi-agent system through the communication and the collaboration of the agents. The system deals with the automated tasks of data processing and handles the different requests from applications and services and manages data from the heterogeneous data sources.

The System is composed of the following components:

- Receiver agent: this agent represents the interface of the system with the data sources. The receiver agent gets data of different types and from different sources like sensors, external databases, web services, etc., and at different rates in streaming or by batch.

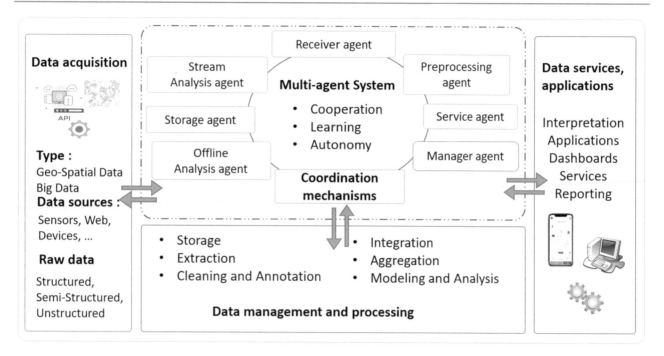

Fig. 6 General overview of the functionalities of the proposed multi-agent system

- Storage agent: this agent is responsible for operations related to storage management and reading data from the Hadoop cluster.
- Offline analysis agent: This agent is responsible for the operations of processing data by batch; this type of analysis can be done on any time, and it can be planned to deal with analysis using historical data, stored in the Hadoop cluster. Batch processing can help in automating tasks of periodical data analysis.
- Stream analysis agent: the role of this agent is to deal with processing the flow of data generally in real time, the use of technologies that generate data in real time like sensors and devices that send requests and data in real time. Those requests wait for responses from the system immediately. The Stream analysis agent in collaboration with the offline analysis agent can solve many type of processing requests like those they need in the same time; processing data in the same time based on the historical data already stored, this collaboration can be done through the manager agent.
- Data transformation agent: this agent is capable of changing the format of data based on the processing needs, to facilitate the learning phase by applying operations of data normalization, aggregation and filtering.
- Data reduction agent: this agent plays an important role to reduce time latency, by reducing the data, and produces representative data with a reduced volume, in this phase various technique exists such as clustering, sampling and data compression.

- Service agent: this agent is able to communicate with the applications and services and transfer the results of the processing, through interfaces and user-friendly applications. Through this agent, the users can also plan or execute some tasks.

4.2 The Proposed Multi-agent System Workflow

The operating principle of the proposed multi-agent system is described in Fig. 7; it contains the necessary steps to guarantee the functioning of the system in different situations, through the coordination of the work between agents that compose the system.

The workflow has two entry points, the first from the data sources and the second is the service and applications, and describe the functionality of the system in different scenarios. For example, the system can initiate the processing of the data based on a request received from a sensor or a device, and applies multiple tasks that require the collaboration of all the components of this system. This concept can be applied in several applications such as its use as a recommender system in real time. When the request pass from the receiver agent to the manager, it will be dispatched into tasks according to the type of data processing required. The tasks of storing and processing data will be handled by the pre-processing agent, the online analysis and offline analysis. The tasks related to respond to real time queries based on the

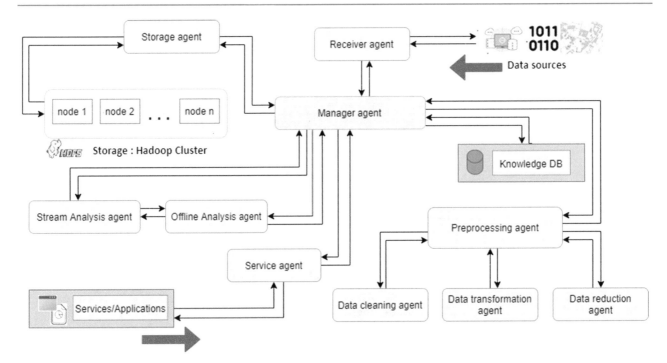

Fig. 7 Agent's coordination workflow

historical data will use also require the use of the storage agent. And the response of the query can be transferred by the service agent.

- In the case of a request from data sources, the receiver agent receives the request and attempts to process it according to the data type, and sends it to the manager. The manager agent who coordinates the work of the different other agents decides whether to store it directly by sending it to storage agent or to send it to the processing agent to pre-process this data using the data cleaning, transformation and reduction agents. After the pre-processing phase, the agent manager decides, depending on the rate of the data received and the requests to be answered, to send this data to the data processing agents: the stream analysis which processes the data in real time and the offline analysis which processes the requests based on batch processing or on historical data. After completion of the processing steps, the agent manager sends the result or the requested data to the application and service layer through the agent service.
- In the case of a request from the service and application layer as well as in the case of planning a reporting task or a particular processing or request to be executed. The agent manager sends the result or data requested through the agent service to the application and service layer, this request goes through the agent service, which transmits the request to the agent manager. The Manager can collect the necessary data from the data cluster or collect the

data from the data sources; and depending on the type of this request decides to send this data to one of the data processing agents, in stream or offline analysis. Then performs the necessary pre-processing through the pre-processing agents and returns the result through the agent service.

5 Tests and Results

In this section, we will study the operation of a micro functionality of our system, the feature responsible for storage analysis and data batch processing. This part studies the use of datasets already collected and stored in the storage cluster, and then the application of different operations like data transformation, pre-processing. The learning phase is based on a machine learning model; this model will be used in the context to help in the decision-making process in smart cities and to serve in urban management and planning, and to predict the different types of habitats according to the characteristics of the population.

Efficient urban planning can help to provide the necessities and needs of the citizens by taking decisions in long and medium term in order to deal with the challenges of managing cities, especially cities facing population explosion density like Casablanca. The proposed solution can be helpful to take the right decisions for urban planning and taking long-term decisions for the city and the different districts by proposing the adequate distribution for the facilities and equipment.

5.1 Data Set: Individuals, Households

For the learning and testing of the model, we used the dataset that contains information about individuals and households in Morocco. Specifically in the city of Casablanca, these data are from the latest census (General Census of Population and Housing 2014, and the data are published in public from 2019); the dataset contains anonymous micro data that includes a sample of 10% of all units except nomadic households, homeless people and the population counted separately.

This dataset is considered sufficient for a better representativeness of the population [31] and to study the population according to the main demographic and socio-economic characteristics such as gender, place of residence, age groups, type of housing, ...

In the experimentation part for each habitat type, we distinguish between seven habitat types in the census data:

T1: Villa/Floor of villa
T2: Apartment
T3: Moroccan House
T4: Summary house/Slum
T5: Rural housing
T6: Other
T7: Not determined.

5.2 Results and Metrics

In the learning phase of a supervised model for classification, we pass through several stages; in our context, we will study the scenario based on learning from a large dataset already recorded and stored in our data storage cluster.

This scenario of processing and analysis of historical data, recorded as a batch of data, will be retrieved, and then proceed to the preparation step; in this step, we will perform pre-processing tasks necessary to the data, such as data cleaning to deal with missing or aberrant data. We performed also some additional pre-processing operations like data transformation and normalization.

After the data pre-processing phase, we move on to the partitioning of the data by dividing it into two partitions: training datasets, which in our case will contain 80% of the data and the test data, and 20% of the test data to evaluate the accuracy of our model.

The objective in this part is to build a classification model capable of predicting the type of housing of the inhabitants based on their socio-economic information such as age, place of residence, level of education, income, ...

For this purpose, it is necessary to try to build the classification model suitable for the type of data being studied using several classification algorithms. Several algorithms have been tried and the three that have given more results have been presented.

For each phase of learning and then creating a new model, an evaluation step is necessary to evaluate the model and adjust its parameter settings to have the best settings for a given algorithm. After the correct hyper-tuning of the model comes a step of saving and then using the model for future prediction queries (Fig. 8).

In the purpose of choosing the best model based on a supervised learning algorithm in our study, we proceeded by the creation, the training and the parameter setting of various models to choose the most suitable model in this case study. The next step is to choose the best model, in this step based on the metrics used in classification problems, like the accuracy, precision, recall and f1-score; we can have an idea about the accuracy and the performance of our model.

In this step, the model with the high metrics like the accuracy will be chosen as a model. Based on this approach, the chosen model will be the best-tested model, with the best parameter tuning (Fig. 9).

The next results will show the details of the metrics of each model:

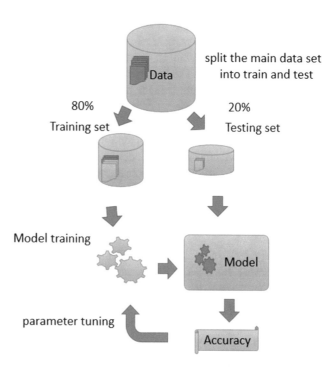

Fig. 8 Machine learning model using classification algorithm

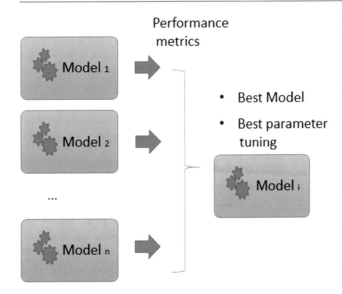

Fig. 9 Selection of machine learning algorithm

- Multi-layer Perceptron (Table 1).

- Gaussian Naïve Bayes (Table 2).

- Random Forest Algorithm (Table 3).

The tables shown previously presented the main metrics used to study the accuracy of a model; the results are shown for every class, which represent a type of habitat.

According to the previous results presented in the tables, which calculates the metrics for each algorithm used in this study, we have the classifier based on the Multi-layer perceptron with an accuracy of 73%, the model based on Gaussian Naïve Bayes with an accuracy of 61% and the model based on Random Forest Algorithm with an accuracy of 85%. In this study, the Model based on the Random Forest Algorithm has the best results.

Based on the ability of the model to predict and have a best accuracy, we will opt for the model based on random forest algorithm which is the best model; in our case, the best results are given by the Random Forest algorithm.

5.3 Data Visualization

The step of data visualization is an important phase, which tries to present the results of the data processing in a user-friendly interface or a comprehensive report to facilitate the task of getting valuable information from the process of data management and analytics to the end user. In our study, after the steps of data storage, data pre-processing, data processing, building a machine learning model, we used this model to predict the type of habitat for the population of the

region Casablanca-Settat. The result of this prediction can be exported using csv files or directly in a database. That information can be shown on a map for every province in this region with the help geographic system like QGIS or ArcGIS.

The map shows the type of habitat with proportions in each province; for example, in Casablanca, apartment is the most demanded type of habitat; in Berrchid and Eljadida Moroccan, house is the most appeared; and the province Sidi Bennour known for agriculture activities has rural housing as the most common type of habitat (Fig. 10).

The results presented in the map can provide a basis on which decisions can be made to adjust housing policy in the Casablanca-State region.

For example, it can be observed that the dominant type of housing in Casablanca city is apartments, and this is normal given the large number of middle class people living in this city. Which may prompt decision-makers to consider that city planning should take into consideration, the preparation of housing type areas in the right proportions for each type of housing and make the necessary provisions for the various facilities required such as the number of schools and hospitals.

6 Conclusion

In this paper, we had proposed a multi-agents model for automating big data analysis and processing based on the smart data approach in order to support urban planning and management and enhance the decision-making process. A state of art was presented where we cited the concept and paradigms in relation to the studied problem.

We presented also some related works to suit our work. In the section of the proposed model, we explained the functionalities and the workflow of the agents, and we presented a case study of the allocation of machine learning algorithms for classification to support the decision-making.

The proposed approach is also extensible; we can add other agents if needed, due to the powerful paradigm of distributed artificial intelligence of the multi-agent system.

The use of the proposed model can help Automating Urban planning and management in smart cities, and support decision-making in addition to other points:

- In the majority of cases, there is an absence in the automation of tasks, especially in the area of decision-making and data science.
- The automation of such process will have a beneficial effect on organizations and companies.
- To increase their return on investment by benefiting from the recommendations, and help for real-time decision-making.

Table 1 Classification report for the model based on MLP

	Precision	Recall	F1-Score
T1	0.68	0.43	0.53
T2	0.81	0.57	0.67
T3	0.68	0.89	0.77
T4	0.89	0.75	0.81
T5	0.00	0.00	0.00
T6	0.00	0.00	0.00
T7	0.00	0.00	0.00
Accuracy			0.73

Table 2 Classification report for the model based on Naïve Bayes

	Precision	Recall	F1-Score
T1	0.29	0.25	0.27
T2	0.57	0.79	0.66
T3	0.74	0.52	0.61
T4	0.76	0.66	0.70
T5	0.03	0.16	0.05
T6	0.09	0.11	0.10
T7	0.06	0.21	0.09
Accuracy			0.61

Table 3 Classification report for the model based on Random Forest

	Precision	Recall	F1-Score
T1	0.94	0.57	0.71
T2	0.83	0.83	0.83
T3	0.83	0.89	0.86
T4	0.95	0.91	0.93
T5	0.96	0.45	0.61
T6	0.99	0.24	0.38
T7	1.00	0.77	0.87
Accuracy			0.85

- Optimize the production and monitoring costs of the observed or studied process; this optimization will have a positive effect on the optimization of the income on investment compared to such a type of system.
- The automation of such a process in a Big Data environment will contribute to the resolution of a number of issues related to the storage, use, analysis and exploitation of advanced techniques of Big Data Analytics to manage real-time events, and to provide a set of KPIs allowing real-time monitoring.
- Favoring autonomous systems based on knowledge bases that contain expertise in a given domain, and on machine learning techniques, this use can also contribute to the decrease of the factor related to human error in decision-making in some situations.

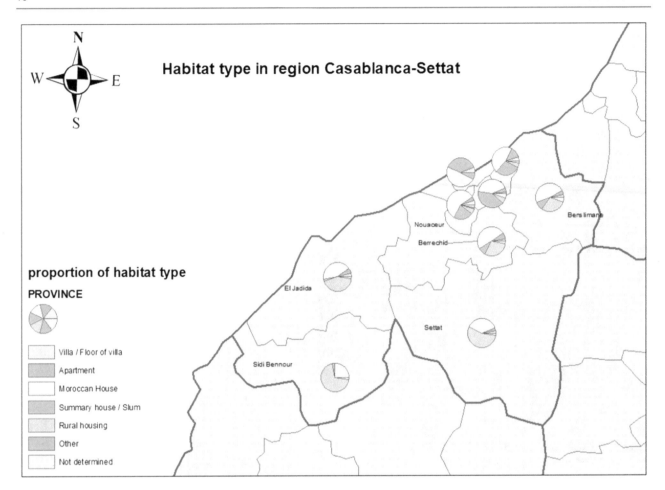

Fig. 10 Proportion of habitat type in region Casablanca-Settat

References

1. C. Liang, L. Zhou, Research on distributed storage of big data based on HBase Remote Sensing Image, in *2019 IEEE 4th Advanced Information Technology, Electronic and Automation Control Conference (IAEAC)*, Dec. 2019, vol. 1, pp. 2628–2632. https://doi.org/10.1109/IAEAC47372.2019.8998001
2. E. Dumbill, Making sense of big data. Big Data **1**(1), 1–2 (2013). https://doi.org/10.1089/big.2012.1503
3. Y. Qin, Q.Z. Sheng, N.J.G. Falkner, S. Dustdar, H. Wang, A.V. Vasilakos, When things matter: a survey on data-centric internet of things. J. Netw. Comput. Appl. **64**, 137–153 (2016). https://doi.org/10.1016/j.jnca.2015.12.016
4. P. Ni, C. Zhang, Y. Ji, A hybrid method for short-term sensor data forecasting in Internet of Things, in *2014 11th International Conference on Fuzzy Systems and Knowledge Discovery (FSKD)*, Aug. 2014, pp. 369–373. https://doi.org/10.1109/FSKD.2014.6980862
5. S.N. Odaudu, I.J. Umoh, M.B. Mu'azu, E.A. Adedokun, Machine learning for strategic urban planning, in *2019 2nd International Conference of the IEEE Nigeria Computer Chapter (NigeriaComputConf)*, Oct. 2019, pp. 1–7. https://doi.org/10.1109/NigeriaComputConf45974.2019.8949665
6. W.-W. Choi, J.-W. Ahn, D.-B. Shin, Study on the development of Geo-Spatial Big Data service system based on 7V in Korea.
7. KSCE J. Civ. Eng. **23**(1), 388–399 (2019). https://doi.org/10.1007/s12205-018-1764-1
7. S.-C. Yu, D.-B. Shin, J.-W. Ahn, A study on concepts and utilization of Geo-Spatial Big Data in South Korea. KSCE J. Civ. Eng. **20**(7), 2893–2901 (2016). https://doi.org/10.1007/s12205-016-0504-7
8. J. Luengo, D. García-Gil, S. Ramírez-Gallego, S. García, F. Herrera, Smart data, in *Big Data Preprocessing: Enabling Smart Data*. ed. by J. Luengo, D. García-Gil, S. Ramírez-Gallego, S. García, F. Herrera (Springer International Publishing, Cham, 2020), pp. 45–51
9. A. Lenk, L. Bonorden, A. Hellmanns, N. Roedder, S. Jaehnichen, Towards a taxonomy of standards in smart data, in *2015 IEEE International Conference on Big Data (Big Data)*, Santa Clara, CA, USA, Oct. 2015, pp. 1749–1754. https://doi.org/10.1109/BigData.2015.7363946
10. C. Bishop, *Pattern Recognition and Machine Learning* (Springer, New York, 2006)
11. M.S. Mahdavinejad, M. Rezvan, M. Barekatain, P. Adibi, P. Barnaghi, A.P. Sheth, Machine learning for internet of things data analysis: a survey. Digital Commun. Netw. **4**(3), 161–175 (2018). https://doi.org/10.1016/j.dcan.2017.10.002
12. A. Dey, Machine learning algorithms: a review. (IJCSIT) Int. J. Comput. Sci. Inf. Technol. **7**(3), 1174–1179 (2016)
13. X. Zhu, A.B. Goldberg, Introduction to semi-supervised learning. Syn. Lect. Arti. Intell. Machine Learn. **3**(1), 1–130 (2009). https://doi.org/10.2200/S00196ED1V01Y200906AIM006

14. P.P. Shinde, S. Shah, A review of machine learning and deep learning applications, in *2018 Fourth International Conference on Computing Communication Control and Automation (ICCUBEA)*, Aug. 2018, pp. 1–6. https://doi.org/10.1109/ICCUBEA.2018.8697857

15. J. Döllner, Geospatial artificial intelligence: potentials of machine learning for 3D point clouds and geospatial digital twins. PFG **88** (1), 15–24 (2020). https://doi.org/10.1007/s41064-020-00102-3

16. S.-H. Zhang, N. Gu, J.-X. Lian, S.-H. Li, Workflow process mining based on machine learning, in *Proceedings of the 2003 International Conference on Machine Learning and Cybernetics (IEEE Cat. No.03EX693)*, Nov. 2003, vol. 4, pp. 2319–2323. https://doi.org/10.1109/ICMLC.2003.1259895

17. L. Breiman, Random forests. Mach. Learn. **45**(1), 5–32 (2001). https://doi.org/10.1023/A:1010933404324

18. P. Subramaniam, M.J. Kaur, Review of security in mobile edge computing with deep learning, in *2019 Advances in Science and Engineering Technology International Conferences (ASET)*, Dubai, United Arab Emirates, Mar. 2019, pp. 1–5. https://doi.org/10.1109/ICASET.2019.8714349

19. X. Zhai, A.A.S. Ali, A. Amira, F. Bensaali, MLP neural network based gas classification system on Zynq SoC. IEEE Access **4**, 8138–8146 (2016). https://doi.org/10.1109/ACCESS.2016.2619181

20. S. Mohammadian, A. Karsaz, Y.M. Roshan, A comparative analysis of classification algorithms in diabetic retinopathy screening, in *2017 7th International Conference on Computer and Knowledge Engineering (ICCKE)*, Oct. 2017, pp. 84–89. https://doi.org/10.1109/ICCKE.2017.8167934

21. M.J. Wooldridge, *An Introduction to Multiagent Systems*, 2nd edn. (Wiley, Chichester, U.K, 2009)

22. E. Belghache, J. George, M. Gleizes, Towards an adaptive multi-agent system for dynamic big data analytics, in *2016 Intl IEEE Conferences on Ubiquitous Intelligence Computing, Advanced and Trusted Computing, Scalable Computing and Communications, Cloud and Big Data Computing, Internet of People, and Smart World Congress (UIC/ATC/ScalCom/CBDCom/IoP/SmartWorld)*, Jul. 2016, pp. 753–758. https://doi.org/10.1109/UIC-ATC-ScalCom-CBDCom-IoP-SmartWorld.2016.0121

23. A. Gandomi, M. Haider, Beyond the hype: big data concepts, methods, and analytics. Int. J. Inf. Manage. **35**(2), 137–144 (2015). https://doi.org/10.1016/j.ijinfomgt.2014.10.007

24. K. Krishnan, *Data warehousing in the age of big data*. Amsterdam [u.a.: Elsevier [u.a., 2013

25. B.T.G.S. Kumara, I. Paik, J. Zhang, T.H.A.S. Siriweera, K.R.C. Koswatte, Ontology-based workflow generation for intelligent big data analytics, in *2015 IEEE International Conference on Web Services*, New York, NY, USA, Jun. 2015, pp. 495–502. https://doi.org/10.1109/ICWS.2015.72

26. A. Shashwat, D. Kumar, A service identification model for service oriented architecture, in *2017 3rd International Conference on Computational Intelligence & Communication Technology (CICT)*, Ghaziabad, India, Feb. 2017, pp. 1–5. https://doi.org/10.1109/CIACT.2017.7977299

27. G. Lombardo, P. Fornacciari, M. Mordonini, M. Tomaiuolo, A. Poggi, A multi-agent architecture for data analysis. Future Int. **11** (2), Art. no. 2 (2019). https://doi.org/10.3390/fi11020049

28. G. Dudek, M.R.M. Jenkin, E. Milios, D. Wilkes, A taxonomy for multi-agent robotics. Auton. Robot **3**(4), 375–397 (1996). https://doi.org/10.1007/BF00240651

29. K. Dounya, K. Okba, S. Hamza, B. Omar, Design and implementation of a new approach using multi-agent system for security in big data. IJSEIA **11**(9), 1–14 (2017). https://doi.org/10.14257/ijseia.2017.11.9.01

30. F. Sassite, M. Addou, F. Barramou, A smart data approach for Spatial Big Data analytics, in *2020 IEEE International conference of Moroccan Geomatics (Morgeo)*, May 2020, pp. 1–6. https://doi.org/10.1109/Morgeo49228.2020.9121920

31. RGPH 2014 | Téléchargements | Site institutionnel du Haut-Commissariat au Plan du Royaume du Maroc. https://www.hcp.ma/downloads/RGPH-2014_t17441.html. Accessed 28 Nov 2020

Geospatial Forecasting and Social Media Exploration Based on Sentiment Analysis: Application to Flood Forecasting

Sara Abas and Malika Addou

Abstract

The objective of this article was to propose a geospatial forecasting approach to analyze the textual content of social media using geospatial components and Natural Language Processing (NLP) tools. This approach has been applied to flood forecasting to mitigate future risks and flood damage since the dynamics of real-world events such as floods prompt users to discuss the topic. The approach is based on the appropriate filtering and preprocessing of information from the Twitter exchange platform. A textual preprocessing method and a new sentiment analysis method (also called opinion extraction method) have been developed to gradually build the database provided by humans. Another method was associated with sentiment analysis, the polarity method which was developed to identify the good or bad feeling attributed to a word in a sentence (positive or negative polarity). The work of this paper offers a different approach of geospatial flood prediction. The method relies on social behavior displayed and made available by users of the platform Twitter. A sentiment analysis method is developed in order to identify users' reactions to inundations according to their geospatial location, while considering how users randomly interact on social media. This paper is innovative compared to the existing approaches, for it uses a social network to extract geospatial components, pairs them with a new data preprocessing and a sentiment analysis method, in order to predict floods on a map, by selecting relevant data. The system thus proposed in this exploration on social media allows the generation of flood forecasting maps to aid the alerting decision-making process. The flood occurrence probabilities provided by this system allow the simulation of the flood forecast distribution map for each month of the year.

Keywords

Geospatial forecasting • Flood forecasting • Social media exploration • Sentiment analysis

1 Introduction

The retrieval and analysis of geospatial data for scientific research is an important part of geospatial scientific research, but also for other fields that benefit from its contribution. Spatial analysis helps solve complex problems and gain a better understanding of what is happening in the world, and where it is happening. It goes beyond simple cartography, allowing the study of the characteristics of places and the relationship between them. Spatial analysis opens up new perspectives in decision-making processes. One of the most intriguing and remarkable aspects of GIS (Geographic Information Systems) [1] is spatial analysis. With spatial analysis, we can combine information from many independent sources and generate an entirely new set of information (results), by applying an elaborate set of spatial operators. This comprehensive set of spatial analysis tools helps us find answers to complex questions in this area. Statistical analysis helps determine whether the phenomena or trends we observe are significant. One of the most useful interactional platforms for gathering such statistics are social networks. Social networks make it possible to observe phenomena according to their spatio-temporal attributes. These observations can be used for several purposes, such as generating analysis of rumors, the medical field, or natural disasters such as inundations.

Urban inundations are posing a threat to economical and humanitarian sustainability. It costs billions of dollars to undo the damage post urban flooding. Hydrodynamic models for determining flood inundation are presented in flooding maps using GIS. The tools developed for several years from GIS are receiving unprecedented echo. Maps

S. Abas (✉) · M. Addou
Architecture, System and Networks Team (ASYR) - Laboratory of Systems Engineering (LaGeS), Hassania School of Public Works EHTP, Casablanca, Morocco

© The Author(s), under exclusive license to Springer Nature Switzerland AG 2022
F. Barramou et al. (eds.), *Geospatial Intelligence*, Advances in Science, Technology & Innovation,
https://doi.org/10.1007/978-3-030-80458-9_2

monitoring and simulation tools find their true place in communication around floods, both in prevention, alerting, and monitoring the event. However, high spatial and temporal resolution data have been challenging to collect because of: certain weather conditions, surveying is very limited in accuracy and disciplined frequency, and climate change. The climate change alone, indicates that historical data used from years ago can lead to invalid results in regards to future predictions. Weather is constantly changing on the planet for millions of years, and is rapid in certain intervals of time such as the current one we are witnessing. Real life dynamic data is the best database for recent updates. Social media is a dynamic platform constantly fed by real-life information provided by the users. This can represent a great database of reference for studying phenomena such as urban flooding. With the rise of technology in the millennial era, communities' self-expression is set free and available to the public. This has led to a widely available data to analyze and study provided by the real-world communities. In the real world, certain natural occurrences and hazards, like environmental disasters, translate to a propagation of information on this topic, on social networks. The equivalent of human action on social media is the information diffusion between the users. This exchange can provide a statement, a warning to flee a geographic area, or even a solution or cautious prevention suggestion to help rescue people in an emergency related event. This paper uses the availability of recent years' user-provided activity on the platform Twitter in order to develop a geospatial method which optimizes the flood risk management, by preprocessing human inserted data, for the purpose of determining a flood prediction using the components provided by social networks related to geospatial and temporal dimensions. The second section of the paper discusses the literature regarding flood prediction, and sentiment analysis. The third section details our approach and the conclusion at last to sum up our work.

2 State of the Art

2.1 Geospatial Aspect

When a person views a map, they subconsciously begin to transform it into information and explore its contents to uncover trends or make decisions based on what they see. Geospatial data designates information related to objects or elements present in a geographic space. Technological advances allow us to capture more spatial data than before. The increasing scale and complexity of data analysis problems require tighter integration of interactive geospatial data visualization with statistical data-mining algorithms [2]. Visualizing large geospatial data sets involves mapping the

two geographical dimensions to screen coordinates and encoding the statistical value by color [3].

Flood forecasting techniques are found in literature using GIS. GIS is a computer system that analyzes and displays geographically referenced information. It uses data that is attached to a unique location. These systems store, analyze, and visualize data for geographic positions on Earth's surface. An example of the use of GIS in flood forecasting and inundation mapping for the Indus River [4]. A river routing model has been developed that can predict the flood condition (discharges and stages) at known locations. Digital maps of the entire Indus Basin are produced in order to show expected flood prone areas. The methodology relies on datasets like 1500 scale maps, satellite imagery, and flood plain maps. Another work [5] uses as the main method a Web-based GIS software as a decision support system for flood events is presented. GIS systems rely on real-life historical databases and satellite provided imagery to introduce flood forecasting methods. An alternative to the use of such databases are real-life reports on virtual platforms such as social media.

2.2 Geospatial Forecasting and Social Media

Social media can also provide a real-life database of geospatial information and events from an alternative perspective than that of the main GIS databases of references. A dynamic database is fed with human-reported events and stories, which can be analyzed and preprocessed using NLP tools to build a consistent relevant database used as the main reference for analysis. The NLP is a field of artificial intelligence (AI) whose main purpose is to develop algorithms that can interpret human language, understand the content of textual statements, and draw conclusions. To achieve that, subtasks of NLP tackle the process of human–machine language. Available NLP platforms are deployed to run NLP oriented algorithms. Most commonly used are: NLTK (Natural Language Toolkit), spaCy [6], Stanford's NLP tools, including the part-of-speech (POS) tagger, the named entity recognizer (NER), the parser, and other provided components.

Social media data can be used for tracking emergency events such as natural catastrophe. They offer a relevant reliable source of data provision for textual and visual analysis. Twitter is the platform of reference used in [7]'s work to track flood phase transitions. The method consists of image classification/topic filtering and a textual geolocating process using an NLP deep learning model, by using an NER tool NeuroNER [8]. This helps identify place names such as "Houston", mostly cities and towns. GeoNames [9] and TIGER road [10] data are used to identify the geospatial data referred to in a textual content. Another work [11] used

NLP techniques to data collected from Twitter, and NER to identify and detect geographic data within user-provided content. Keyword mining is the most used method of reference when it comes to filtering flood-related relevant tweets [12–18]. The process of geolocating using text-mining and NER tools to identify geospatial data can be efficient. Keyword mining is also a good method to use in order to identify the spectrum of flood lexical field. However, these main attributes research aspires to look for and use in order to draw conclusions can lead to error if the topic is not related to inundations regardless of contrary literal evidence. Twitter data is a human-generated dynamic database. The flood keywords might be detected by the programs, however, the topic might have nothing to do with the literal definition of a flood found in a dictionary. In order to remedy this issue, this paper suggests a method to identify sentiment behind flood-related tweets. When a sentiment is positive, it is estimated in this paper that it is less likely related to an actual flood which is the main target of this paper. Sentiment analysis is an important methodology used in this work. As mentioned in the paper [19], a lot of research has been done on sentiment analysis using the twitter platform [20, 21].

Comparison discussion with similar approaches

A lot fewer works have been done for sentiment analysis over spatial perspective, which is the idea of this paper. The paper [19] focuses on developing a geospatial sentiment analysis method as well, to study the Brexit event. In that approach, the author combines geospatial components data and sentiment analysis as well. However, the difference is that the global appreciation is calculated by the average sum of reactions with regard to "Brexit", whether they are positive or negative. The method [19] is statistical, and observational of past occurring events. The method of this paper is probabilistic, and predictive of future events. Another difference between the two approaches is that the present work filters the context before considering the *Tweet* or shared

content. If the content is irrelevant to the context even though it mentions the word of interest, it could lead the results into error. In this paper, we suggest a method to make that differentiation, and it is not manual. As a concrete example, there is another geospatial sentiment analysis method [22] used to identify a tie between sentiment and geospatial temporal effect. For example, weekend events and friend and family gatherings are the time that users prefer to post positive tweets. In the western part of the US, users love to post photos on Twitter more than in other parts of the US. Much like method [19], approach [22] does not filter context, predict, or identify sentiment behind informal words, which are points addressed by this paper. Table 1 compares the different approaches discussed above. The colons represent the methods used in their works, as criteria for comparison.

– ***Keyword Filtering Method***: The operation of searching for relevant tweets using the proper keywords.
– ***Name Entity Recognition (N.E.R)***: The use of NLP libraries to extract entities.
– ***Sentiment Analysis Method (S.A)***: The identification of user sentiment behind their statements.

Formal Language: The formal language referred to is the one found in common knowledge bases used as a reference to evaluate a formal word.

Informal Language: The language used by users on social networks. A lot of times it is invented by the users.

– ***Relevant Tweet Selection Automatic Method (R.T.S.A. M)***: The extraction of tweets relevant to theme of our research, which is in this paper the context of "inundation" as in water floods.
– ***Polarity Quantification Method (P.Q)***: The quantification of the detected polarities or the evaluated sentiment behind global and relevant textual data.
– ***Twitter Geospatial attributes***: The geospatial components made accessible via Twitter Developer tools.

Table 1 Methods used for content analysis and geospatial components retrieval on social media

Approach	Keyword filtering	N.E.R	S.A (informal language)	S.A (formal language)	R.T.S.A.M	P.Q	Twitter geospatial attributes
[4]		✓					
[8]		✓					
[9]	✓						
[19]				✓		✓	✓
[22]				✓			✓
Our approach	✓	✓	✓	✓	✓	✓	✓

2.3 Sentiment Analysis for Social Media and Geospatial Research

The rich amount of user-provided data is a gift from social media to content analysis and geospatial information access. On social media platforms such as Twitter, users express their sentiment when sharing content, and geospatial information is a present component which can be deduced from it. Establishing a statistical function between the geospatial components and the sentiment behind textual content is an innovative method to analyze user feedback by geolocation. User feedback represents user sentiment. Sentiment analysis, also known as opinion mining, is the study of feelings associated to dematerialized textual data. Review sites are a good example of data made available for research purposes. They present a good understanding of the clients' reception perspective of the products and services. They are used as a data source for sentiment analysis models trainings and tests. In the available literature, several opinion-mining studies investigated different approaches. An example of an unsupervised technique for sentiment analysis is Turney's Algorithm [23]. To identify the semantic orientation of a phrase noted as *SO(phrase)*, the author uses Alta Vista [24] as a Web Search Engine resource. It is defined as follows: The positive polarity is calculated by the identification of the phrase with the word "excellent". Similarly, the negative polarity is calculated by the identification of a phrase with the word "poor".

$$SO(phrase) = hits(phrase\ NEAR\ ``excellent'')\ hits(``poor'')$$
$$\log 2\ hits(phrase\ NEAR\ ``poor'')\ hits(``excellent'')$$
$$(1)$$

The hits(phrase NEAR "excellent") "NEAR" operator in the equation represents the co-occurrence of the phrase and the word excellent, through its quantified number of search results in Alta Vista. This method is successful at a 74% accuracy success rate. However, when it comes to words that have several meanings but are written the same, this method leads to the wrong semantic analysis of words. In this paper, a method which identifies a unique meaning to words in tweets is presented. In Ayush Das's work [25], the paper aims to determine the polarity of phrases, and views Turney's Algorithm's method as not independent since it relies on an external search engine. To solve this, the author used SentiWordNet [26] as a resource for sentiment analysis. SentiWordNet contains opinion information related to words extracted from the WordNet database and made publicly available for research purposes. SentiWordNet is built via a semi-supervised method and could be considered a valuable resource for performing opinion-mining tasks. The author uses sentiment weights derived from a word lexicon annotated

with its corresponding polarities. The method proved itself efficient at a 75% accuracy rate. However, nowadays on social platforms like Twitter, new words are constantly emerging. The social network's users come up with new terms to exchange, which don't have a meaning prior to their first emergence on the platform. These words are not available on WordNet, which SentiWordNet relies on in the first place. This work, presents a solution to these newly invented words, which are not available on external database resources. The commonly used sentiment analysis algorithms are Baseline Algorithm [27], Modified Baseline [28], and Turney's Algorithm [29] which were discussed above. Their performance rates vary. However, they are strongly dependent on formal phrasing and they don't take into consideration disambiguation. Other works [30, 31] strongly support the analysis of emoticons used, because they are emotionally expressive. An emoticon is a symbol or icon used, to express an emotion without words. The authors' works' relied on a list of pre-classified emoticons, and emotionally intense words. These approaches have proven to be efficient, however, on platforms like Twitter, emoticons are no longer inserted via text, they are images. A study [30] using available Dutch Tweets, identifies emoticon polarities based on tweets. However, it is focused on text-inserted emoticons. Text-inserted emoticons are icons that show up when the user types certain text sequences. With the updated phone operating systems nowadays, users tweet using the available emojis on their phones. Sentiment analysis is used in this paper in order to identify information which is more likely to truly refer to the inundation lexical field. When a statement is positively polarized, it is less likely to be about a real flood. The following tweet is an example: "I received a flood of congratulations today. Thank you to everyone who did. I am so happy and grateful. Love to all". The word "flood" is detected, however, the polarity of tweet is positive. Given the natural negative polarization humans use to refer to a flood, a positive polarity is not an indicator that the tweet is about a real flood, unless it is the case when it comes to flood recovery. In the case of flood recovery, users tend to tweet less about the previously occurring flood, since it is no longer a trending topic. It is in the phase of disappearance. This is the reason why a negative polarity is considered more associated to a real-life flood.

3 Our Approach

Our approach aims to provide the system's user with a ranked flood likelihood in terms of geographic location as well as month of the inundation occurrence. In order to start the data analysis, the user of the system must start by selecting the preferred language of data. This step is important in order to extract the right lexical field, which is

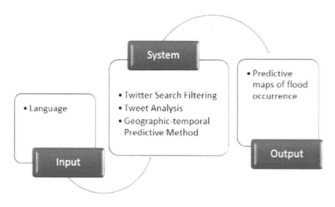

Fig. 1 Method of flood maps probability generation

applied to the flood lexical field in this case. After the proper language is chosen, comes the Twitter Search Filtering step. In this step, the keywords are used in order to retrieve the relevant tweets. In the Tweet Analysis phase, the contribution of the present paper is highlighted the most. This phase is composed of preprocessing, Part-Of-Speech tagging, a sentiment analysis method, and the polarity quantification method. Finally, a geographic-temporal predictive method is suggested to generate predictive maps of flood occurrence, according to each month of the year. These are discussed in detail in the next section. Figure 1 sums up the method on a macroscopic level.

3.1 Twitter Search Filtering

In order to explain the Twitter Search operation, we use the English language as an example. The set of key words that belong to the same semantic field as the word "flood" used to find the Tweets we are looking for is the following one:

$$Keywords = \left\{ \begin{array}{l} deluge\,; inundation\,; tsunami\,; floodwaters\,; rainstorm\,inundated\,; \\ floodwater\,; submerged \end{array} \right\}$$

This set of words is searched for. However, only a tweet that contains at least one of these words, which also expresses a negative sentiment is taken into consideration. A negative sentiment is one that is similar to pain, fear, shock, depression, or disappointment. When a statement reflects this type of emotion, it is said to have a negative polarity. Since this paper focuses on the natural disaster, only negatively polarized tweets are subject to analysis.

We use Twitter Advanced Search option to filter tweets which are shared posts, using the following options:

- *Date of the tweet*: No more than 5-year-old content
  The reason 5-year-old content is chosen is to avoid irrelevancy of an updated situation.
- *Number of votes*: At least 300 likes

When a shared content gets at least 300 votes, this means it is most likely credible.
- *Option of Search*: "Any of These Words"
  Any of the words that belong to the keyword set K.

• **Geospatial components extraction**

First, the selected tweets after the first phase are extracted accordingly to their geographic information. There are two classes of geographic metadata: tweet location, and account location. Given that most users of the platform don't individually geotag their tweets [32], we use the account location when extracting tweets. Geotagging is the act of precising one's location when sharing content. Gnip PowerTrack provides many ways to filter on these types of geographic metadata. It provides an optional Profile Geo enrichment that formalizes the data in the profile location, and makes it more convenient to filter. All Profile Geo coordinates are provided in the [Longitude, Latitude] order. For example, if the user's profile location is: "Denver, CO" With Profile Geo Information enabled, profile location results in setting an array of locations called *user.derived. locations* attribute, which includes: country_code, locality, sub_region, full_name ("Denver, Colorado, United States"), and coordinates: [−104.9847, 39.73915]. Helpful Geooperators are made available such as bio_contains. It performs a substring match on the user's account-level bio. This can help us find users who introduce themselves by mentioning their home location. Another method consists of using Twitter Advanced Search, and combining it with APIs such as Tweepy, or ready-to-use data providing services such as Octoparse. When a user u1 shares a Tweet that we have previously selected, the geolocation of the account is attached to the userID, in case another Tweet by the same user is presented.

Example: $Position(u1) = "Florida, USA"$.

3.2 Tweet Analysis

After having collected the proper tweets related to the flood lexical scope, accordingly to their geographic locations, comes the step of tweet analysis. Figure 2 displays the process used in order to analyze a tweet all the way to quantifying the polarities. The first phase consists of preprocessing the data as a stepping stone to accessing clean data. The second phase is part-of-speech tagging in order to identify the words that can be subject to sentiment analysis; in this case, these words are adjectives as they are the most emotionally expressive ones in a sentence. The third phase is a sentiment analysis method applied to both commonly used

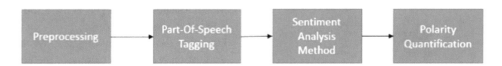

Fig. 2 Global method

words, and unidentified words. The last step in Tweet analysis is polarity quantification where a global feedback is reported as statistics.

Preprocessing

The preprocessing phase is an optimizing important phase in NLP tasks. In this step, a phrase is transformed in order to be interpreted by machine learning algorithms. There are important steps to perform in this NLP subtask. These components consist of:

– Noise removal: replacing irrelevant characters like numbers and special characters with ' '.
– Lowercasing: replacing capital letters with their lowercase sizes.
– Tokenization: Converting word sequences to a String list of Tokens.

Example: *"This overwhelming flood, made us :/ reflect on how important infrastructure is"*

After tokenization and the previous phases, this statement is transformed into the following sequence.

Example: $\begin{bmatrix} \textit{'this', 'overwhelming', 'flood', 'made', 'us', 'reflect', 'on', 'how',} \\ \textit{'important' 'infrastructure', 'is'} \end{bmatrix}$

– Removal of Stopwords: Removal of the common stopwords which are not of use in interpretation such as: "a", "is", "the", "to". In our method, "the" and "a" are not removed because they are used in the next step in order to identify a past-simple verb as an adjective.

After removal of the stopword, the example is transformed to the following list of String characters.

Example: $\begin{bmatrix} \textit{'overwhelming', 'flood', 'made', 'reflect',} \\ \textit{'important' 'infrastructure'} \end{bmatrix}$

– Stemming and Lemmatization [33] are often used in preprocessing. They are not referred to in our method because the aim is to identify adjectives. The NLTK library [34] is used to accomplish these tasks.

Part-of-Speech Tagging

Before performing sentiment analysis, identifying the adjective in the Tweets is one of the steps of our final program. Adjectives are the most emotionally expressive words formulated by a user of a social media platform in a sentence. In order to optimize the method, the sentiment analysis step targets mainly adjectives of the sentence. Libraries such as Stanford NLP [35] or online tools such as Part-of-speech.info [36] are used in order to determine the adjectives in a sentence.

Example: *"The use of geospatial tools is valuable"*

After using the Part-Of-Speech Tagger, the results should be as follows:

– *tools* → *"Noun"*
– *valuable* → *"Adjective"*

Sentiment Analysis Method

After preprocessing and identifying tags in a sequenced set of words, comes the sentiment analysis phase. The method suggested by this paper (see Fig. 3) consists of analyzing two different types of identified adjectives. When an adjective is commonly used and found in the database used in our method, and when it is only used within the social media platform Twitter. The proposed method addresses the use of commonly known adjectives, new adjectives, and emoticons.

• Extraction of Phrase Adjectives

In this phase, the adjectives identified by the previously explained phase "Part-Of-Speech Tagging" are extracted and are considered representatives of the entire statement's polarities.

Fig. 3 Sentiment analysis method

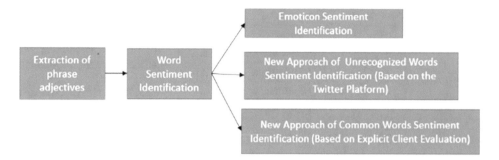

- Word Sentiment Identification

New Approach of Common Words Sentiment Identification

Websites like www.imdb.com (movie reviews), www.amazon.com (product reviews), www.yelp.com (restaurant reviews), www.CNET download.com (product reviews), and www.reviewcentre.com, host millions of product reviews by consumers. For the sample use, dataset for research purposes is made available by these platforms, such as Amazon. This includes reviews (ratings, text, helpfulness votes), product metadata (descriptions, category information, price, brand, and image features), and links (also viewed/also bought graphs).

After extracting the selected tweets accordingly to their geographic attributes, comes the step of polarity detection. In order to optimize the computation of our method, only adjectives are taken into consideration. An adjective is more emotionally descriptive of mentioned information. We use Stanford's NLP part-of-speech (POS) tagger tool, to identify the adjectives.

Figure 4 explains the process of assigning a value to an adjective found in the relevant data that the system analyzes. The number of stars reflects the emotion behind the adjective used to describe the client experience. "Tweet 1" is a phrase subject to analysis. This is how the process of identifying the sentiment and therefore the polarity behind happy works in this article.

To fluidly explain the method, let's take an example. Our program should find similar results: Let the following be

tweets about the flood event extracted, identified by id on the platform.

Examples: *"The deluge is disastrous."* (*Tw1, area : Atlantic city*)
"A tsunami of people showed up to my wedding. I am so happy." (*Tw2, area : Florida*)

The user who tweeted Tw1, refers to the deluge as "overwhelming" which is detected by the Part-Of-Speech tagger as an adjective. In our dataset sample, we select reviews which contain the label "overwhelming". The reason behind choosing labels instead of the expressed paragraphs is for optimization purposes. Let's say that in our sample we have four reviews, as displayed in Table 2.

These reviews are paired as such with their ratings: (Review 1, 0 stars), (Review 2, 1 star), (Review 3, 2 stars), (Review 4, 1 stars).

The variable $ak(w1)$ is used to define whether a review is associated to more than 3 stars or less.

$$ak(w1) = Sk \tag{2}$$

Sk is a binary variable that takes the following values.

- If the review Rk has 3 or more stars, $Sk = 1$.
- Otherwise, $Sk = 0$.

$Mv(w1)$ is the matching vector of the word $w1$ in a set of n reviews from a dataset. All n reviews including the word $w1$. The colons are the reviews.

$$Mv(w1) = (a_1(w1)a_2(w1) \ldots a_n(w1)) \tag{3}$$

Fig. 4 Word-based database exploration method

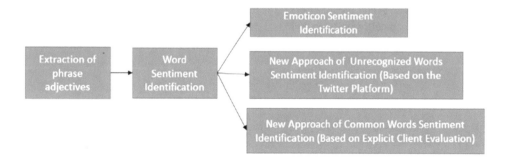

Table 2 Examples of review

Review	Statement
1	This is disastrous. Bad product
2	I don't like it
3	Needs improvement
4	I don't enjoy it

The polarity of the word $w1$ is then defined as such:

$$P(w1) = \frac{\sum ak}{n} \qquad (4)$$

- If $P(wj) \geq \frac{1}{2}$; then the word wj has a positive polarity.
- If $P(wj) = \frac{1}{2}$; then the word wj is neutral.
- Otherwise, the polarity of the word is negative.

Now that we developed the equation of polarity, let's move on to the phrase's polarity. A phrase contains a set of words, like verbs, nouns, adjectives, adverbs. We identify the adjectives using the Part-Of-Speech tagger. If the user used adjectives to describe their experience along with an emoticon, the system would assume that the polarity of the emoticon is that of the adjective.

Emoticon Sentiment Identification

In case the user does not use an adjective but uses an emoticon, we use a sample of 1000 random tweets containing the used emoticons. The choice of 1000 is made solely for optimization of computation. If the font of the editor cannot display the character or if it detects it as an image, it is copied from its original form, then searched for on Twitter Advanced Search Tool. The dominant polarity of adjectives detected within those tweets, using the method above, sets its polarity.

New Approach of Unrecognized Words Sentiment Identification

Certain words that are identified by the Part-Of-Speech Tagger as adjectives can not be found on Amazon databases much less on official knowledge bases such as SentiWordnet [26]. In order to correctly identify the sentiment behind these words, the main database reference would be the platform itself. Users of Twitter can start using an adjective they collectively agree on. When such word is detected, in other words, when it is not found in the preferred reviews database of choice, its polarity is identified by the polarity of the adjectives and emoticons it coexists with.

Example: "*Flood is disastrous. People are bummed out :/*"

The polarity of the word "bummed" is the sum of polarity of adjective "disastrous" and the polarity of the emoji referred to with ":/". Since *Polarity(phrase k)* = −1, with phrase being the Tweet minus the unusual adjective, the polarity of bummed is negative. This task is applied on a sample of 500 phrases containing the same word "bummed". The following equation is the average polarity found in sentences when the unusual adjective occurs.

$$Avg\_Polarity(wk) = \frac{\sum polarity(phrase\ i)}{500} \qquad (6)$$

- If $Avg\_Polarity(wk) \geq \frac{1}{2}$; then the phrase has a positive polarity, and *Polarity(wk)* = *1*.
- If $Avg\_Polarity(wk) = \frac{1}{2}$; then the phrase is neutral, Polarity(*wk*) = 0.
- Otherwise, the polarity of the phrase is negative, *Polarity (wk)* = −1.

The polarity of the word is then added to the polarities of words and emoticons of the tweet of desire.

Polarity Quantification

The purpose of this phase is to assign a polarity to a whole statement expressed by a user. This means a single polarity (negative, or positive) is assigned to a phrase. A statement is said to have a positive polarity if it aligns with the words people use to refer to emotions similar to happiness, joy, pride, gratitude, peace. It is negative if, as mentioned before, it refers to sentiments similar to sadness, disappointment, fear, or anger.

If the phrase contains an adjective: Let m be the number of identified adjectives. wk is a detected adjective in the phrase, while bk is a binary variable

$$Po(Phrase) = \frac{\sum bk}{m} \qquad (5)$$

- If $P(wk)$ is positive then $bk = 1$. If $P(wk)$ is negative then $bk = 0$.
- If $Po(Phrase) \geq \frac{1}{2}$; then the phrase has a positive polarity, and $Polarity(phrase) = 1$.
- If $Po(phrase) = \frac{1}{2}$; then the phrase is neutral, Polarity $(phrase) = 0$.
- Otherwise, the polarity of the phrase is negative, $Polarity$ $(phrase) = -1$.

If the polarity is negative, then the tweet is acknowledged and counted via an incrementation of x_{ij}, j referring to the month of its publication, and i refers to its geographic location.

Relevant Data Selection

Only negative polarity detected tweets are selected, because the method aims to find posts that refer to an inundation event which is not spoken of metaphorically. If we were to use the method [19] by calculating the average of sentiment expressed with regard to inundations, by considering all the tweets, the method would not be efficient. Given a fixed month of the year, and a fixed geospatial area, if most Tweets containing inundation keywords refer to it positively, and few report it in a negative sentiment, the average would be positive. In this case, we would judge the area as less likely to be inundated, even though it isn't true.

We use the following method to automatically disqualify Tweets containing inundation keywords and a positive sentiment at the same time, for the method is looking to identify natural flood disaster, and therefore a negative sentiment.

Table 3 explains the process of associating Tweets to their sentiment, geographic, and temporal data. When the polarity detected is positive, the flood is most likely not the one referring to literal inundations. In this case, there is no extraction of the geographic location or the temporal component for algorithm optimization purposes.

3.3 Geographic-Temporal Predictive Method

In order to find out the probability rank of geographic and temporal co-occurrence of a flood, a sentiment analysis method, as well as the following matrix are dynamically developed if the polarity of the phrase is negative. The matrix below represents an example for three geographic areas, and 3 months.

The variable xij refers to the quantified number of times tweets have a negative polarity in area i and month j.

$$M(English) = \left\{ \begin{array}{ccc} +5 & \dots & +1 \\ +13 & xij & +2 \\ +15 & \dots & +3 \end{array} \right\} \tag{7}$$

This matrix is just an example. It is in reality multidimensional with 12 colons and contains as many lines as geospatial areas identified. It can be translated to Table 4, another example of five area during 3 months for further explanation. From a sample of 500 filtered Tweets, each time a Tweet is referenced in a negative sentiment, the variable is incremented. If the sentiment is positive or neutral, it is completely omitted. X_{2k} represents the variable of line 2; colon k.

Every colon represents a month of the year. The global matrix is composed of 12 colons, and n ligns. n represents the number of Tweets detected referring the flood concept in a negative way. Every line represents a geographic location identified from a selected Tweet's user's account geolocation or bio description attributes provided by the Twitter API as mentioned in the previous section. Every time the month corresponding to the colon j occurs in the location corresponding to the line i while the polarity is negative, the variable x_{ij} is incremented. Let (m, n) be the final obtained dimensions of the matrix. The most incremented variable (which represents the maximum in Eq. (8)) obtained in a fixed area, during a fixed month is considered the 100% chance reference. The occurrence probability of a flood within a precise month is calculated by the following equation:

$$Flood\_Probability(Zone\,i, Month\,k)$$
$$= \frac{xik * 100}{\max(x1k, x2k, x3k\dots, xmk)} \tag{8}$$

Table 3 Tweet characteristics and components retrieval	Tweet	Expressed sentiment	Geographic zone	Temporal component
	I had a **flood** of compliments today. I'm so **happy :D**	Positive	Not extracted	Not extracted
	The **flood** has been quite the **handful**. It's been **tiresome**	Negative	Mississippi	03 December 2019
	Flood is **big**. It was shockingly **fast**, it's **sad**. It was **devastating**	Negative	Florida	22 January 2020

Table 4 Assignment matrix illustration

Areas	January	February	March
Florida	+30	+25	+7
Louisiana	+20	X_{2k}	+3
California	+15	+17	+3
New York	+10	+14	+2
New Jersey	+5	+10	+2

3.4 Results and Validation

Table 5 is the application of the formula (8). It represents the output results for five different geographic areas in the month of January. For the month of January, the mathematical results of the calculated flood probabilities, as previously presented in Table 4 from a Tweet sample, are presented in Table 5.

Figure 5 is an example of a map that would illustrate the month of January for statistics made available on www. statista.com/statistics. In this case, our probabilistic formula represents a highly similar map, for the percentages shown in Table 5, which are based on a Twitter sample, are similar to the ones noted in the areas. The result of our system are 12 predictive maps of flood occurrence, one for each month. For the purpose of better explanation, only the month of January was the subject of numerical study.

A statistical map is a cartographic representation on which quantitative data are represented (see Fig. 5). It is used to represent the value of a statistical variable in each of the geographic units of a global entity. It is a data visualization tool. The statistical map has the advantage of being able to both reveal a global analysis while allowing everyone to locate details for each geographic unit. A heat map is a data visualization technique that shows magnitude of a phenomenon as color in two dimensions. Softwares like StoryMap JS (Knight Lab) [37] and ArcGIS [38] are tools to help the user tell stories on the web that highlight the locations of a series of events. Carto [39] is also a great tool for data visualization. Below is our application with an illustration, which displays the inundation probabilities for the month of January in the USA area.

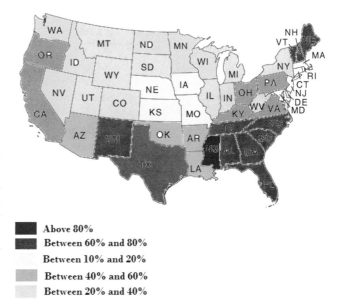

Color	Legend
■	Above 80%
■	Between 60% and 80%
□	Between 10% and 20%
▨	Between 40% and 60%
▢	Between 20% and 40%

Fig. 5 Predictive flood map of the month of January in USA

4 Conclusion

In this paper, a geospatial flood-forecasting approach was proposed to contribute to flood risk management techniques using NLP-based methods. It consisted of analyzing social media textual content from the Twitter platform, with the use of the geospatial components and NLP tools in order to preprocess data. This paper proposes a unique method, which consists of combining geospatial component extraction methods and sentiment analysis to predict floods. It is different from other methods because a new approach is elaborated to omit irrelevant data, detect sentiment of formal as well as informal words, and the proposed formula translates the obtained results without a survey intervention. A new method for preprocessing textual data is also developed. Probabilities of flood occurrence were calculated after proper analysis of user generated data related to inundations. The calculated estimations of flood occurrence accordingly to their geospatial and temporal components are represented with flood prediction maps.

Table 5 Probability distribution for the month of January

Geographic zone	Flood probability formula results (%)
Florida	100
Louisiana	67
California	50
New York	30
New Jersey	17

References

1. M. Mohamed, R. Plante, Remote sensing and geographic information systems (GIS) for developing countries, in *International Geoscience and Remote Sensing Symposium*, IEEE, Toronto, Ontario, Canada (2002), pp. 2285–2287

2. D.A. Keim, C. Panse, M. Sips, S.C. North, Visual data mining in large geospatial point sets. IEEE Comput. Graphics Appl. **24**(5), 36–44 (2004)

3. P. Bak, M. Schäfer, Density equalizing distortion of large geographic point sets. First publ, in *Cartography and Geographic Information Science*, pp. 237–250 (2009)

4. H. Sharif, M.A. Hashmi, Use of RS & GIS in flood forecasting and early warning system for Indus Basin, in *International Conference on Advances in Space Technologies*, Islamabad (2006), pp. 21–24

5. D. Mioc, L. Gengsheng, F. Anton, B.G. Nickerson, Decision support for flood event prediction and monitoring, in *IEEE International Geoscience and Remote Sensing Symposium*, Barcelona (2007), pp. 2439–2442

6. https://spacy.io/

7. R. Wang, Y. Hu, Z. Zhou, K. Yang, Tracking flooding phase transitions and establishing a passive hotline with ai-enabled social media data. IEEE Access **8**, 103395–103404 (2020)

8. F. Dernoncourt, J.Y. Lee, P. Szolovits, NeuroNER: an easy-to-use program for named-entity recognition based on neural networks (2017)

9. https://www.geonames.org/

10. W. Song, T. Haithcoat, J. Keller, A snake-based approach for TIGER road data conflation, in *Cartography and Geographic Information Science* (2006)

11. R.Q. Wang, H. Mao, Y. Wang, C. Rae, W. Shaw, Hyper-resolution monitoring of urban flooding with social media and crowdsourcing data. Comput. Geosci. J. **111**, 139–147 (2018)

12. D. Laney, 3D data management: controlling data vol. velocity and variety. META Group Res. Note **6**(70), 1 (2001)

13. Z. Li, C. Wang, C.T. Emrich, A novel approach to leveraging social media for rapid flood mapping: a case study of the 2015 South Carolina floods. Cartography Geographic Inf. Sci. J. 97–110 (2015)

14. M.F. Goodchild, J.A. Glennon, Crowdsourcing geographic information for disaster response: a research frontier. Int. J. Digital Earth **3**, 231–241 (2010)

15. A. Sheth, Citizen sensing, social signals, and enriching human experience. IEEE Int. Comput. (2009)

16. M. Nagarajan, A. Sheth, S. Velmurugan, Citizen sensor data mining, social media analytics and development centric web applications, in *Proceedings of the 20th International Conference Companion on World Wide Web*, New York, NY, USA (2011), pp. 289–290

17. N.R. Adam, B. Shafiq, R. Staffin, Spatial computing and social media in the context of disaster management. IEEE Intell. Syst. 90–96 (2012)

18. J. Fohringer, D. Dransch, H. Kreibich, K. Schröter, Social media as an information source for rapid flood inundation mapping. Nat. Hazards Earth Syst. Sci. **15**, 2725–2738 (2015)

19. A. Agarwal, R. Durga, Geospatial sentiment analysis using twitter data for UK-EU referendum. J. Inf. Optim. Sci. **39**, 1–15 (2018)

20. M. Daniel, R.F. Neves, N. Horta, Company event popularity for financial markets using Twitter and sentiment analysis. Expert Syst. Appl. J. **71**, 111–124 (2017)

21. L.Y.F. Su, M.A. Cacciatore, X. Liang, D. Brossard, D.A. Scheufele, M.A. Xenos, Analyzing public sentiments online: combining human-and computer-based content analysis. Inf. Commun. Soc. **20**(3), 406–427 (2017)

22. T. Hu, B. She, L. Duan, H. Yue, J. Clunis, A systematic spatial and temporal sentiment analysis on geo-tweets. IEEE Access **8**, 8658–8667 (2020)

23. P.D. Turney, Thumbs up or thumbs down?: semantic orientation applied to unsupervised classification of reviews, in *Proceedings of the 40th Annual Meeting on Association for Computational Linguistics* (2002), pp. 417–424

24. https://search.yahoo.com/?fr=altavista

25. A. Das, Sentiment analysis, in *8th International Conference on Computing, Communication and Networking Technologies* (2017)

26. S. Baccianella, A. Esuli, F. Sebastiani, SentiWordNet 3.0: an enhanced lexical resource for sentiment analysis and opinion mining, in *Proceedings of the International Conference on Language Resources and Evaluation*, Valletta, Malta (2010)

27. B. Pang, L. Lee, S. Vaithyanathan, Thumbs up sentiment classification using machine learning techniques. Department of Computer Science Cornell University Ithaca, NY 14853 USA and IBM Al Maden Research Centre 650 Harry Rd. San Jose, CA 95120 USA (2002)

28. S. Das, M. Chen, Yahoo! for amazon: extracting market sentiment from stock message boards, in *8th Asia Pacific Finance Association Annual Conference, APFA* (2001)

29. G.S. Solakidis, K.N. Vavliakis, P.A. Mitkas, Multilingual sentiment analysis using emoticons and keywords, in *IEEE/WIC/ACM International Joint Conferences on Web Intelligence (WI) and Intelligent Agent Technologies* (2014)

30. D.B. Hogenboom, F. Frasincar, Exploiting emoticons in sentiment analysis, in *Proceedings of the 28th Annual ACM symposium on Applied Computing* (2013), pp. 703–710

31. J. Read, Using emoticons to reduce dependency in machine learning techniques for sentiment classification, in *Proceedings of the ACL Student Research Workshop*. Association for Computational Linguistics, Stroudsburg, PA, USA (2005), pp. 43–48

32. L. Sloan, J. Morgan, Who tweets with their location? Understanding the relationship between demographic characteristics and the use of geoservices and geotagging on Twitter. PLOS ONE J. **10** (11) (2015)

33. I. Boban, A. Doko, S. Gotovac, Sentence retrieval using stemming and lemmatization with different length of the queries. Adv. Sci. Technol. Eng. Syst. J. **5**, 349–354 (2020)

34. B. Wagh, J. Shinde, P.A. Kale, A Twitter sentiment analysis using NLTK and machine learning techniques. Int. J. Emerg. Res. Manage. Technol. **6**, 12 (2018)

35. C. Manning, M. Surdeanu, J. Bauer, J. Finkel, S. Bethard, D. McClosky, The Stanford Core NLP natural language processing toolkit, in *Proceedings of 52nd Annual Meeting of the Association for Computational Linguistics: System Demonstrations* (2014)

36. https://parts-of-speech.info

37. https://storymap.knightlab.com/

38. https://www.arcgis.com/index.html

39. https://carto.com

Deep Convolution Neural Network for Automated Method of Road Extraction on Aerial Imagery

Norelyaqine Abderrahim, Abderrahim Saadane, and Azmi Rida

Abstract

The detection of roads from satellite images is among the most important topics for planning and development for cities, which replaces manual methods, but they turn out to be a complex task due to the complexity of the objects. This paper discusses issues related to the detection and segmentation of roads in very high-resolution aerial images. In order to resolve these issues, we propose in this paper the use, deployment, and validation of deep learning strategies, in particular, the U-net architecture based on deep convolutional neural networks for extracting roads from remote sensing images. Data augmentation techniques and preprocessing were applied to improve accuracy. We use in our processing for road segmentation the Massachusetts Road Dataset, which is publicly available. The results obtained showed excellent performance in terms of recall, precision, accuracy, and F1 score, and they are very close to the ground truth; it outperforms all other models presented, with a high accuracy of 97.7%.

Keywords

Remote sensing • Deep learning • Road extraction • Image classification

N. Abderrahim (✉)
Laboratoire de géophysique appliquée, de géotechnique, de géologie de l'ingénieur et de l'environnement (L3GIE), Mohammed V University – Mohammadia School of Engineering, Rabat, Morocco

A. Saadane
Department of Geology, Faculty of Sciences of Rabat, University Mohammed V, Rabat, Morocco

A. Rida
Center of Urban Systems, Mohamed VI Polytechnic University (UM6P), Ben Guerir, Morocco
e-mail: rida.azmi@um6p.ma

1 Introduction

Remote sensing is a scientific discipline that combines a wide range of skills and technologies, generally used to remotely acquire information about terrestrial objects without making physical contact, using the properties of the electromagnetic waves emitted by these objects.

This technique is very important and sometimes crucial for observing and understanding the living environment: weather forecasts, land monitoring (vegetation mapping, changes in agriculture, changes in cities, etc.), military surveillance, and study of the evolution of the oceans. Since their appearance, digital sensors in the 1970s have been used to provide images with adequate spatial and spectral resolutions to understand the phenomenon of urban expansion. They hence have become real planning and development tools for cities, especially in rapidly changing areas where up-to-date spatial information is needed.

Indeed, observation of the urban environment from space has taken off since the appearance of civilian satellite images with spatial resolution. This resolution is still increasing, tending toward a resolution of less than one meter. The quality of detail visible on such images has aroused great interest among experts involved in the study of the urban environment. The increase in spatial resolution also increases the quantity of data generated. However, the manual processing of the huge mass of data present in high spatial resolution satellite images is costly and time-consuming.

The demand for new automatic data processing techniques (remote sensing images) became more and more pressing since the extraction of objects in these images is useful for many applications. Recognizing these objects and evaluating their spatial positions and relationships is considered a pattern recognition problem. Among the information sought in remote sensing images, road networks are of particular interest because of the variety of their applications. These elements seem to be overriding in geographic information systems. However, their extraction is delicate and much of

© The Author(s), under exclusive license to Springer Nature Switzerland AG 2022
F. Barramou et al. (eds.), *Geospatial Intelligence*, Advances in Science, Technology & Innovation,
https://doi.org/10.1007/978-3-030-80458-9_3

the work has been consecrated to the study of this issue. The detection of road networks from a satellite and aerial image proves to be a complex task due to the complexity of the objects. This can be explained by the nature of the observed network. Indeed, networks have very different appearances according to their types (motorway network, road network, paths…), their contexts (rural, peri-urban, urban, or forest), or their dates of construction. An urban network in a large city, for example, is likely to appear as a network with a grid structure, whereas winding roads in the countryside will have a much less defined structure. Furthermore, with the recent availability of very high spatial resolution satellite imagery, we are able to locate the road and extract it as a surface feature more accurately. On the other hand, it puts us in front of the complexity of the objects, caused by (trees along the streets, vehicles…). Indeed, road mapping is still a major challenge for computer algorithms, whereas for human interpreters it is an immediate task.

Road detection from satellite images captures many researchers for more than two decades. Early work on road detection used specific operators that measure the degree of belonging to a road for each pixel, by calculating its neighbors, such as the DUDA operator described by Fischler et al. [1]. Roux et al. [2] proposed another improved version of this operation. Shen and Castan [3] which used the infinite symmetrical exponential filters (ISEF) to extract roof and valley profiles in satellite images. Some studies [4] consider the analysis of the longitudinal variance of the road, and this type of method works for roads with homogeneous radiometry and good contrast with the environment; however, roads do not always have homogeneous radiometry, especially in an urban environment. For road extraction Fischler et al. [1] proposed a dynamic programming method. Mathematical morphology has been widely used in remote sensing, for example, Roux et al. [2] apply (CHF) operator to extract intensity peaks in the SPOT image; nevertheless, the CHF is not very selective and gives noisy results. For road extraction in high-resolution images Zhang et al. [5] use the concept of granularity, and the method remains sensitive to the problem of partial occlusion due to the presence of buildings near the road or tree shadows that give a discontinuous appearance of the road. Saradjian and Amini [6] show that satellite images can be simplified using mathematical morphology operators and propose two structuring elements for road vectorization. Cai et al. [7] present an approach based on region growth by machine learning and apply it to automatic road detection from satellite images.

State of the art on road extraction was presented by Mena [8] for GIS updating from satellite images and aerial photographs, and a literature review of nearly 250 references is presented. Among the semi-automatic methods, active contour algorithms are used by Laptev et al. [9], and the marked point processes underlying stochastic geometry and reversible jump MCMC dynamics are applied to satellite images and aerial photographs by Stoica et al. [10].

The methods of learning classification played a very important role and have been the subject of several works. Mokhtarzade and Zoej [11] address for road detection in Ikonos and Quickbird images the possibility of using neural networks, and they use fuzzy clusters and genetic algorithms to improve their results. Today, several factors lead us to the use of deep learning techniques for better results. First of all the abundance of images, artificial neural networks, during the training process need a large number of images. This need is met by the multiple data sets of several million labeled images that are available, and with hardware improvements the number of images that can be used in the learning process has increased. Convolutional (CNNs) and fully convolutional neural networks (FCNs) are today very powerful tools for object classification [12]. These networks achieve the best results on most public remote sensing reference databases with performances between 80 and 90% of global accuracy. Rezaee and Zhang [13] used a CNN patch-based to detect roads from images with 1.2 m spatial resolution of Fredericton city, and the model shows more accuracy than SVM.

Based on the recent success of deep learning, and to better extract roads from aerial images with good classification, we have been inspired, in this paper, by the U-net architecture which has been able to surpass many techniques presented in the literature, and this architecture has gained significant popularity and has proven its effectiveness in many medical image segmentation applications.

2 Related Work

2.1 Semantic Segmentation

In computer vision, semantic segmentation is the task defined as assigning a class to each coherent region of an image. This can be achieved in particular by classifying each pixel of the image. There are many methods of segmentation which can be divided into the following:

Contour-Based Segmentation

Contour-based segmentation is concerned with the contours of objects in the image. Most of the algorithms belonging to this segmentation family are local, which means they work at the pixel level. The main problems related to this approach which we must first characterize are the borders between objects and also that the contours are often incomplete, and therefore it is necessary to be able to close them. Many operators exist to detect the contours, for example that of Martin et al. [14]; they have detected image boundaries using local brightness, textures, and colors, and they used as input into a regression model. These methods have now been improved notably by using multi-resolution information.

Region-Based Segmentation

Region-based segmentation approaches use different criteria, such as spatial or temporal criteria to find homogeneous regions. The best-known method of region-based segmentation is watershed segmentation [15]. This method is based on mathematical morphology which considers a grayscale image as a topological relief on which a flood is simulated. Numerous improvements have been made to this algorithm, notably on the placement of water sources, and this type of method is still widely used, notably in remote sensing [16].

Segmentation by classification has evolved enormously over time, especially with the appearance of deep learning. Originally, this type of method was generally used in addition to another method. For example, Song and Civco [17] are the first to use a SVM to roughly segment images into two classes. However, there are also methods using only a machine learning approach such as Bertelli et al. [18] who use the HOG descriptor with an SVM. The use of deep learning for segmentation started in 2009, and more and more studies have focused on convolutional networks to perform segmentation tasks, and the results were very promising. In 2012 Socher et al. [12] proposed to initially use a single convolution layer and pooling. This first step serves as their feature extractor. On this extractor, they stack recursive neural networks (RNNs) which are used on local blocks of the extractor output.

Today, CNN is the benchmark in many applications of the computer community, and several researchers have turned to this type of solution for the problem of object detection, particularly in image classification for remote sensing [13]. Besides, several models are based on convolutional auto-encoders, such as SegNet [19] and U-net [20].

2.2 Deep Learning for Remote Sensing

Automated mapping from remote sensing data with deep learning has been the subject of several recent studies. In particular, several approaches derived from computer vision have been successfully used to remote sensing images. Many of the work using deep learning methods in remote sensing are based solely on the use of optical images. Indeed, these images are closer to the traditional images available in large databases. The CNNs are specifically dedicated to images. They have shown astonishing performances in object recognition and detection. Nonetheless, it was not very used

in the classification of satellite images in remote sensing. Hu et al. [21] use the first CNN layers ImageNet datasets for high-resolution optical image classification functions. The use of such network on natural images has been shown more efficient than classical classification methods. In this way, the approaches have been very successful thanks to changes in image classification provided by CNNs. Several works deploy CNNs for the detection of buildings [22]. The results show significant improvements compared to other traditional classification methods.

Indeed, the models qualified as fully convolutional networks (FCN) obtained excellent results on high-resolution images in urban areas [23], and it was among the first applications of FCN on optical aerial data, based on the initial architectures of Long et al. [24]. Approaches using FCN for semantic segmentation of remote sensing images have become very popular. Indeed, FCNs infer a pixel prediction for the whole image in a single pass, thus overcoming the problem of patch classification, and this drastically reduces computation times. Symmetrical encoder-decoder models quickly follow and are the subject of several derivative works, such as the edge regularization by explicit constraint [25]. Although this work is limited to optical imagery, the fusion of heterogeneous data has also been studied subsequently. In particular, efficient deep neural network architectures using dual-input FCNs have been proposed for multispectral/SAR [26].

2.3 Deep Learning for Road Detection

The detection of the road from satellite or aerial images has been the subject of much studies, and several methods have been proposed. Indeed, the stakes are high since the extent of the areas to be mapped is immense and the time required to update existing maps is considerable. However, despite all the attention paid to the problem, due to the large variability of the objects concerned, road detection is becoming an important issue, and consequently, the difficulty to characterize them. Since Mnih's first work [22] using CNN for the extraction of roads and buildings in aerial images, these approaches have been successfully used on many very high resolution data [27]. And different from the MNIH method, to build deep neural networks using restricted Boltzmann machines, Panboonyuen et al. [28] proposed a CNN directly on aerial imagery of Massachusetts road dataset, and they showed better results than the MNIH model. Zhong et al. [29] proposed fully convolutional networks for road detection from satellite images. Road structure refined CNN was proposed by Wei et al. [30] for road networks, and he shows a remarkable improvement in the results.

3 Methodology

3.1 Dataset Description

We use in our processing the Massachusetts Road Dataset [22], published by the University of Toronto and available online, which contains a total of 1171 aerial images with a spatial image of around 1 m per pixel. The dataset covers an area of over 2600 km². Figure 1 shows the RGB image and its corresponding ground truth classified into two segments (road and no road) from the Massachusetts dataset with a size of 1500 × 1500 pixels. Table 1 describes the dataset for training, testing, and validation.

3.2 Data Augmentation

Today's computing power enables the construction of very large networks of deep neurons with up to hundreds of billions of parameters. The amount of learning data is crucial to estimate such networks. Howard [31] believe that increased data is fundamentally important to improving network performance. Data augmentation involves artificially increasing the size of the learning database by adding new examples created from distortions of the original examples. The objective is that the network learns descriptors specific to the classes of objects considered rather than image artifacts such as differences in illumination. For example, for classical images, it is clear that an object does not change if the ambient lighting changes or if the observer is replaced by another one. Particularly, for satellite images, the presence of an object such as clouds in a satellite image is independent of the rotation of the image. The final model must be less sensitive to the colors that are determined by the illumination conditions of the scene for classical images and to the orientation of the image caused by the acquisition conditions of the satellite for satellite images.

Data augmentation consists of applying certain transformations on the initial learning set to create new artificial examples. The applied transformations do not change the nature of the detected class and thus create new examples. Data augmentation allows to artificially create invariances of the final network and to increase the generalization performance [31]. For classical images, among the classical operations used for data augmentation we recall: rotation, enlargement, and translation as shown in Fig. 2. More advanced transformations such as changing the contrast or brightness of the image can also be added according to the user's needs.

3.3 Network Architecture

Based on convolutional networks, Ronneberger et al. [20] propose a new architecture allowing the use of a reduced number of training data. The main quality of the approach, presented above, is to provide the output spatial information and not only a label of the class to which the image belongs. We associate a label to each pixel of the image. For this, it is not the whole image that is passed as input of the network,

(a)   (b)

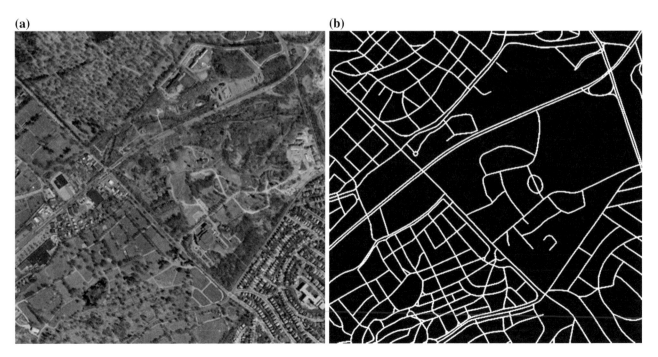

Fig. 1 **a** The RGB image, **b** the corresponding label from Massachusetts road dataset

Table 1 Overview of dataset. Numbers of training, testing, and validation sets

Dataset	Training	Testing	Validation
1171	1108	49	14

Fig. 2 Data augmentation.
a Original satellite image,
b Transformed image

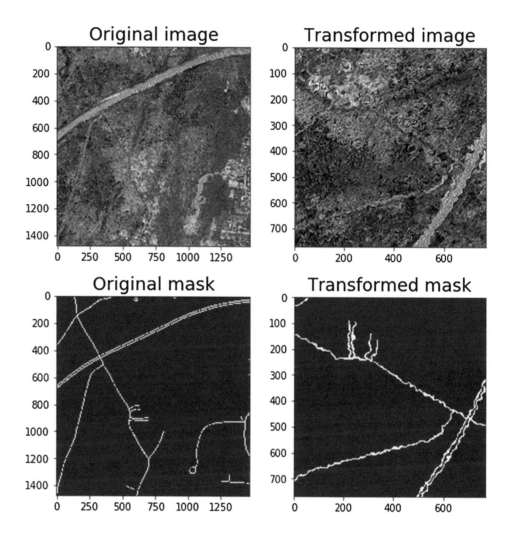

but patches centered in each of the pixels to be classified. This has the effect of multiplying the size of the training data, but at the same time increasing the calculation time. To cross each pixel with overlapping patches, redundant data are involved. Moreover, by proceeding locally, one loses contextual information that could be useful for the final segmentation. There is therefore a trade-off between defining a high size of the patches to keep information on the spatial structure (we then lose precision) and defining a small size of the patches to gain precision (we lose contextual information here).

Seyedhosseini et al. [32] overcome this problem by learning the contextual information of the image hierarchically, with their method "hierarchical waterfall model (CHM)", he illustrated in Fig. 3. On sub-sampled images, a classifier is trained at each level of the hierarchy, as well as on the max-pooling results of the previous steps. The outputs

of the classifiers are then oversampled to the initial resolution of the input image and are used to train a new classifier.

Applied to segmentation, this technique allows precise localization and keeps the spatial context of the image. Nonetheless, it requires training several classifiers. For a number of 328 images (average size 250×250), 35 h of training are required.

Ronneberger et al. [20] then proposes to adapt this approach by using a convolutional network.

The idea is to use the architecture of Seyedhosseini et al. [32] where each classifier can be approached by a layer of a convolutional network. Thus, when entering the network, the image to be segmented is convoluted by 64 filters with a stride of 1, followed by a switch to the rectified linear unit (ReLU) activation function. This step repeated twice. It should be noted that the edges of the images and the maps generated are not managed. After the first steps of

Fig. 3 Cascaded hierarchical
model (CHM)

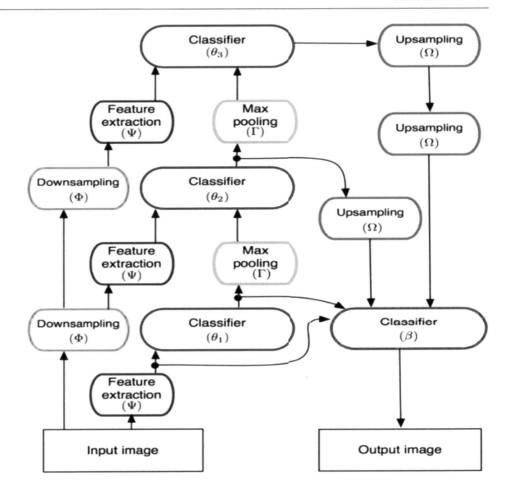

convolution and transition to ReLU, a max-pooling is performed with a stride of two. This has the effect of
sub-sampling the previous data by a factor of two. This is
done until a 32 × 32 size card is obtained.

The data is then over-sampled. Convolution, ReLU and
oversampling iterations are then carried out by propagating
the contextual data until a segmentation map is obtained.
This depth map 2 indicates the position of the pixels and the
associated class. Therefore, we obtain a U-shaped architecture composed of a contraction phase (encoder) then an
expansion phase (decoder). The architecture is shown in
Fig. 4.

The steps of convolutions, ReLU activation, and
max-pooling are described below:

Convolution

A convolution layer is set by a number N of convolution maps,
the size of the convolution kernel (often square), allowing the
detection of specific characteristics, which are applied to the
input images. Each convolution map is sum of convolution
maps of the previous layer by its respective convolution kernel. There are 64 convolutions in the U-net network per layer,
and they have a fixed size of 3 × 3 (see Fig. 5).

Pooling

The principle of pooling consists on obtaining in output map of
property invariant to small translations by combining several
localities of an input property map. The most common pooling
operation is max-pooling, which replaces a rectangular area
with the highest value locality it contains. The major advantage of this operation is to improving the efficiency of the
net-work by avoiding overlearning, which is a limiting factor
in learning a CNN, also to reduce the size of the processed data
in memory. As a general rule, pooling windows are convolved
on the input property map with a step equal to the size of the
convolution window. Therefore, there is no overlap or space
between two convolution slots. We used a max-pooling of size
2 ∗ 2 with a step of 2 as illustrated in Fig. 6.

ReLU Function

The ReLU function is a crucial function in the whole process
used after each convolution. Graphically it looks like Fig. 7.

The advantage of this function, also known as the activation function, is without affecting the receptive fields of
the convolutional layer, and it allows to increase the
non-linear properties of the entire network and the decision

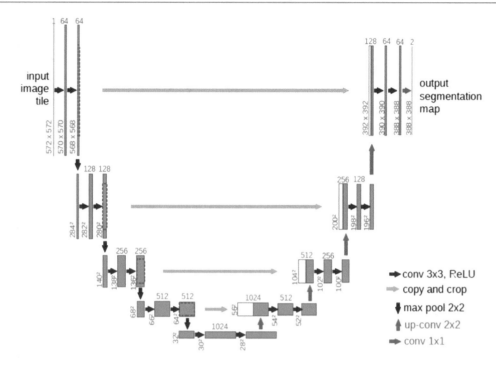

Fig. 4 Original U-net for image segmentation

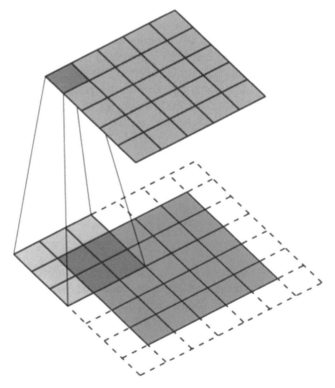

Fig. 5 Convolution operation on an image (convolution layer of size 3 by 3 which scans the input pixels (blue), to obtain an output map (green))

function. The result of a ReLU layer is the same size as what is passed to it as input, with all negative values removed.

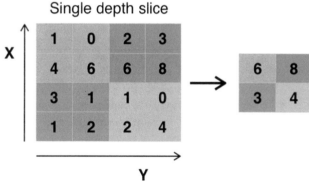

Fig. 6 Pooling operation with a 2×2 filter and a step of 2

4 Experiments and Results

4.1 Metrics for Classification

At last, to evaluate the relative results of the various classification and segmentation models, it is necessary to define quantitative criteria allowing comparison.

We define the following performance metrics for the classifier:

- The precision is defined as the ratio between the number of true positives and the total number of elements assigned to the class by the classifier:

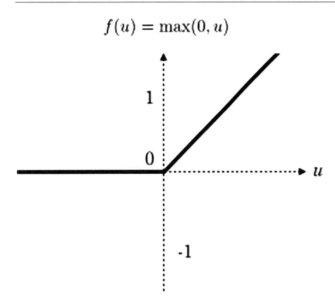

$$f(u) = \max(0, u)$$

Fig. 7 The graph of ReLU. Linear Activation Function for Positive and Negative Inputs

$$\text{Precision} = \frac{\text{TP}}{\text{TP} + \text{FP}} \qquad (1)$$

- The recall is defined as the ratio between the number of true positives and the total number of elements really belonging to the class:

$$\text{Recall} = \frac{\text{TP}}{\text{TP} + \text{FN}} \qquad (2)$$

- The F1 score, or Sorensen-Dice coefficient, is defined as the harmonic mean of precision and recall:

$$\text{F1} = \frac{2 \times \text{Precision} \times \text{Recall}}{\text{Precision} + \text{Recall}} \qquad (3)$$

4.2 Testing and Results

We choose to compare our model with different algorithms based on deep learning for detection roads on the same database "Massachusetts Road dataset" already used in our model. The Recall, Precision, Accuracy, and F-score metrics are shown in Table 2.

It can be seen in Table 2 that our network based on U-Net architecture outperforms all other models; in terms of recall and precision, it works better than all three other approaches, it produces the best Accuracy value with 97.7%, and also he achieved the highest F-score of 87.5%.

We can finally see in Fig. 8, which shows the results obtained presented by a predicted map of the U-net model. The results obtained from the proposed model, as we can observe, resemble several advantages and they are very close to the ground truth, which is encouraging. Our model can segment the roads with high precision and less noise, as it is shown in the first row, it can define two-lane roads well without any problem, and they are well filled and thickened. And also our model produces relevant results in complex structures; it can also successfully distinguish between road structures and similar structures such as parking lots. In addition to that, as we can see in the second row, he can identify roads that are not labeled in the ground truth.

From the results obtained, in order to improve the output of our model, we want to better understand the reason for the loss of precision. Not just the training protocols make the model more or less efficient, we have observed also over the evaluations and predictions made on our data that the model could be sensitive to certain factors, which would directly influence the quality of the prediction. The limit of our model could be its specialization on an architectural type. Indeed, we have noticed that it is easier to detect certain types of roads than others. We observe that the correspondence between the ground truth road and the prediction mask is higher when it comes to the less complex roads. And also another, that there are roads which are not labeled, the masks are badly defined by OSM, and this behavior has been observed on many images, therefore our model considers them as background. And there is always room for improvement, especially in updating satellite images.

5 Conclusion

In this paper, we have listed different ways of segmenting roads. We have mentioned some classical methods which are still very widely used such as morphological methods, and also in recent years more modern methods based on learning which have shown their excellent precision. Much progress has been made in segmentation and visual recognition in

Table 2 Comparison of performances in testing Massachusetts data among different deep learning methods for roads detection

Model	Precision (%)	Recall (%)	Accuracy (%)	F-score (%)
FCN [29]	43.5	68.6	90.4	53.2
RSRCNN [30]	60.6	72.9	92.4	66.2
SegNet [28]	77.3	76.5	95.7	76.8
U-Net	**86.8**	**88.3**	**97.7**	**87.5**

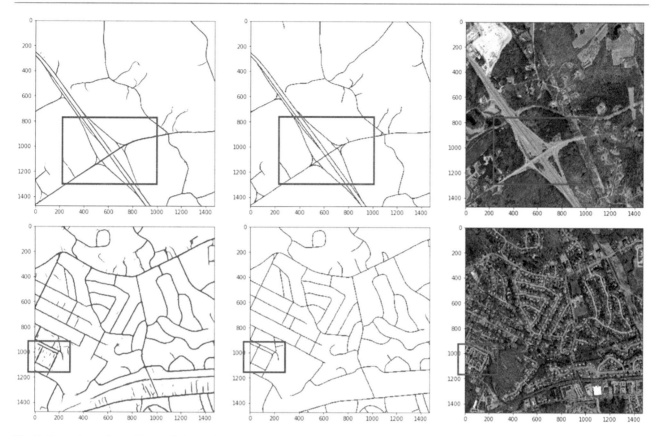

Fig. 8 Example results. **a** The predictions of the proposed U-net, **b** The ground truth, **c** The original image

general with the reappearance of deep neural networks and more particularly of convolutional neural networks. They allow a new approach based on the hierarchical feature extraction of high levels of images. Thus, several promising techniques are born based on these architectures.

We then presented the U-net architecture, which allows efficient road segmentation by quickly training a deep neural network. We implemented an adapted architecture with which we were able to experiment on several images from the "Massachusetts Road dataset" database. The results are very encouraging, and they have surpassed the results of many techniques mentioned previously. However, we came up against certain material limits.

Our work is only in its initial version, and we can say that this work remains open for the improvement of a model which will be able to segment the roads with more precision and using a small number of training samples.

References

1. M.A. Fischler, J.M. Tenenbaum, H.C. Wolf, Detection of roads and linear structures in low-resolution aerial imagery using a multisource knowledge integration technique, in *Readings in Computer Vision* (Elsevier, 1987), pp. 741–752

2. M. Roux, J. Lopez-Krahe, H. Maître, Recalage image SPOT/carte routière. Int. Arch. Photogramm. Remote Sens. **29**, 384–384 (1993)

3. J. Shen, S. Castan, An optimal linear operator for step edge detection. CVGIP: Graph. Mod. Image Process. **54**(2), 112–133 (1992)

4. S. Airault, O. Jamet, Détection et restitution automatique du réseau routier sur des images aériennes. Traitement du signal **12**(2), 189–200 (1995)

5. C. Zhang, S. Murai, E.P. Baltsavias, Road network detection by mathematical morphology, in *ISPRS Workshop 3D Geospatial Data Production: Meeting Application Requirements 1999*. Institute of Geodesy and Photogrammetry, ETH-Hoenggerberg (1999)

6. M. Saradjian, J. Amini, Image map simplification using mathematical morphology. Int. Arch. Photogramm. Remote Sens. **33**, 36–43 (2000)

7. X. Cai, A. Sowmya, J. Trinder, Learning parameter tuning for object extraction, in *Asian Conference on Computer Vision* (Springer, 2006), pp. 868–877

8. J.B. Mena, State of the art on automatic road extraction for GIS update: a novel classification. Pattern Recogn. Lett. **24**(16), 3037–3058 (2003)

9. I. Laptev, H. Mayer, T. Lindeberg, W. Eckstein, C. Steger, A. Baumgartner, Automatic extraction of roads from aerial images based on scale space and snakes. Mach. Vis. Appl. **12**(1), 23–31 (2000)

10. R. Stoica, X. Descombes, J. Zerubia, A Markov point process for road extraction in remote sensed images (2000)

11. M. Mokhtarzade, M.V. Zoej, Road detection from high-resolution satellite images using artificial neural networks. Int. J. Appl. Earth Obs. Geoinf. **9**(1), 32–40 (2007)

12. R. Socher, B. Huval, B. Bath, C.D. Manning, A. Ng, Convolutional-recursive deep learning for 3d object classification. Adv. Neural. Inf. Process. Syst. **25**, 656–664 (2012)

13. M. Rezaee, Y. Zhang, Road detection using deep neural network in high spatial resolution images, in *2017 Joint Urban Remote Sensing Event (JURSE)* (2017, IEEE), pp. 1–4

14. D.R. Martin, C.C. Fowlkes, J. Malik, Learning to detect natural image boundaries using local brightness, color, and texture cues. IEEE Trans. Pattern Anal. Mach. Intell. **26**(5), 530–549 (2004)

15. S. Beucher, Use of watersheds in contour detection, in *Proceedings of the International Workshop on Image Processing* (CCETT, 1979)

16. A. Anand, Brain tumor segmentation using watershed technique and self organizing maps. Indian J. Sci. Technol. **10**(44) (2017)

17. M. Song, D. Civco, Road extraction using SVM and image segmentation. Photogramm. Eng. Remote. Sens. **70**(12), 1365–1371 (2004)

18. L. Bertelli, T. Yu, D. Vu, B. Gokturk, Kernelized structural SVM learning for supervised object segmentation, in *CVPR 2011* (IEEE, 2011), pp. 2153–2160

19. V. Badrinarayanan, A. Kendall, R. Cipolla, Segnet: a deep convolutional encoder-decoder architecture for image segmentation. IEEE Trans. Pattern Anal. Mach. Intell. **39**(12), 2481–2495 (2017)

20. O. Ronneberger, P. Fischer, T. Brox, U-net: convolutional networks for biomedical image segmentation, in *International Conference on Medical Image Computing and Computer-Assisted Intervention* (Springer, 2015), pp. 234–241

21. F. Hu, G.-S. Xia, J. Hu, L. Zhang, Transferring deep convolutional neural networks for the scene classification of high-resolution remote sensing imagery. Remote Sens **7**(11), 14680–14707 (2015)

22. V. Mnih, Machine learning for aerial image labeling. Citeseer (2013)

23. J. Sherrah, Fully convolutional networks for dense semantic labelling of high-resolution aerial imagery (2016), arXiv:1606.02585

24. J. Long, E. Shelhamer, T. Darrell, Fully convolutional networks for semantic segmentation, in *Proceedings of the IEEE Conference on Computer Vision and Pattern Recognition* (2015), pp. 3431–3440

25. D. Marmanis, K. Schindler, J.D. Wegner, S. Galliani, M. Datcu, U. Stilla, Classification with an edge: improving semantic image segmentation with boundary detection. ISPRS J. Photogramm. Remote Sens. **135**, 158–172 (2018)

26. J. Hu, L. Mou, A. Schmitt, X.X. Zhu, FusioNet: a two-stream convolutional neural network for urban scene classification using PolSAR and hyperspectral data, in *2017 Joint Urban Remote Sensing Event (JURSE)* (IEEE, 2017), pp. 1–4

27. J.E. Vargas, P.T. Saito, A.X. Falcao, P.J. De Rezende, J.A. Dos Santos, Superpixel-based interactive classification of very high resolution images, in *2014 27th SIBGRAPI Conference on Graphics, Patterns and Images* (IEEE, 2014), pp. 173–179

28. T. Panboonyuen, K. Jitkajornwanich, S. Lawawirojwong, P. Srestasathiern, P. Vateekul, Road segmentation of remotely-sensed images using deep convolutional neural networks with landscape metrics and conditional random fields. Remote Sens. **9**(7), 680 (2017)

29. Z. Zhong, J. Li, W. Cui, H. Jiang, Fully convolutional networks for building and road extraction: preliminary results, in *2016 IEEE International Geoscience and Remote Sensing Symposium (IGARSS)* (IEEE, 2016), pp. 1591–1594

30. Y. Wei, Z. Wang, M. Xu, Road structure refined CNN for road extraction in aerial image. IEEE Geosci. Remote Sens. Lett. **14**(5), 709–713 (2017)

31. A.G. Howard, Some improvements on deep convolutional neural network based image classification (2013), arXiv:1312.5402

32. M. Seyedhosseini, M. Sajjadi, T. Tasdizen, Image segmentation with cascaded hierarchical models and logistic disjunctive normal networks, in *Proceedings of the IEEE International Conference on Computer Vision* (2013), pp. 2168–2175

Enhancing the Management of Traffic Sequence Following Departure Trajectories

Bikir Abdelmounaime⬤, Idrissi Otmane⬤, and Khalifa Mansouri⬤

Abstract

Nowadays, several busy airports have an issue in sequencing departure aircrafts while respecting the prescribed defined minima by the international organizations (such as the Euro control and International Civil Aviation Organization—ICAO) following their departure trajectories (the standard instrument departures—SIDs or omni-directional trajectories), answering to the order of aircrafts' demands of taxiing and taking off, especially when following the First Come First Served (FCFS) method and handling their respected categories. The purpose of this paper is to establish an algorithm that optimizes the departure sequence taking into account the aircraft categories, the spent time on the taxiway, the runway, and during the climb in the departure trajectory. The used algorithm based on the Shortest Job First (SJF) concept will at first sequence the traffic in the ground according to the estimated time to reach the holding point to involve a better use of the departure trajectories. This work will also offer a comparative study of various numbers of aircrafts and show the gained time in comparison with the FCFS concept.

Keywords

Air Traffic Flow Management • Aircraft sequencing problem • Departure trajectories • Departure sequence optimization

1 Introduction

Air transport is considered to be one of the domains in continuous expansion all over the world. According to Euro control official statistics of 2019 [1], traffic growth was 0.9% with a total of over 11.1 million flights. This expansion from another side is facing a stable or a slow growth in airports infrastructure which generates more pressure on air traffic controllers to handle the demand of air traffic movements, especially in rush hours.

The disproportion between demand and air capacity is the main cause of delays without taking into account measures that can be forced by weather or applied by other technical issues. As mentioned in Fig. 1, it is clear that delays on airports are greater than those in the en-route phase which confirms the wasted time and energy while taxiing and waiting in the holding points.

The cost of increasing the capacity of the traffic network is huge, demanding and related to many constraints in different levels which lead many stakeholders of air traffic management into searching for other solutions and methods so as to optimize traffic flows. Many efforts focused on finding suitable and practical models for the terminal areas and airports grounds as they considered to be the critical phases in terms of lost time, energy, and the difficulty to handle for air traffic controllers, especially in busy airports.

This work will focus on the aircraft sequencing problem (ASP) just after takeoff while following their assigned departures trajectories.

2 Problematic

Aircrafts after takeoff follow the standard instrument departure (SID) or the assigned departure trajectory by air traffic controllers. An SID is a designated instrument flight rule (IFR) departure route linking the aerodrome or a specified runway of the aerodrome with a specified significant

B. Abdelmounaime (✉) · I. Otmane · K. Mansouri
Laboratory Signals, Distributed Systems and Artificial Intelligence
ENSET, University Hassan II, Mohammedia, Morocco

© The Author(s), under exclusive license to Springer Nature Switzerland AG 2022
F. Barramou et al. (eds.), *Geospatial Intelligence*, Advances in Science, Technology & Innovation,
https://doi.org/10.1007/978-3-030-80458-9_4

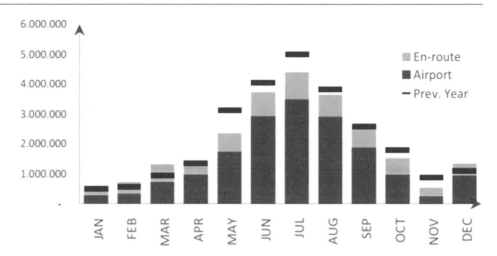

Fig. 1 Air Traffic Flow Management (ATFM) delays in 2019 per month

point, normally on a designated route (Air traffic services—ATS), at which the en-route phase of a flight commences [2].

Airfields that can receive instrument rules flights may dispose of one or many SIDs depending on the available runways and the surrounding space structure (mountains, danger areas, regulated zones, etc.). These SIDs may be strategically separated or independent or just dependent one of another. In case of dependent SIDs, a delay is generated for aircrafts on the ground on the holding point before takeoff because the precedent traffic is occupying the SID which won't be useful until the traffic reaches a certain point, altitude, or flight level.

Aircrafts types and, respectively, their performances have an impact on the use of the SIDs since the slow-moving ones are taking more time than the fast ones to clear the SID. From another side, the aircraft sequence on the ground is also influencing the use of the SIDs because when low-performance aircrafts are preceding those with high performance it takes more time for the whole sequence to take off and of course more delay for most the aircrafts.

From the above, it is clear that the optimization of the traffic sequence on the ground will involve a better use of the SIDs or departure trajectory and contribute to reducing the delay.

3   State of Art

The authors in paper [3] developed a method relaying on Reference Business Trajectories as a source of data to reduce Air Traffic Controller interventions at the tactical level while preserving Air Traffic Flow Management operations. The proposed work aims to sequencing aircrafts on departure at the airports by taking into account the benefits of small time stamp variations in the assigned Calculated Takeoff Time departures and to improve Trajectory-Based Operations

concepts. In the research [4], the authors present a data-splitting algorithm to ideally solve the aircraft sequencing problem (ASP) which was constructed as a mixed-integer program (MIP), considering several realistic constraints, including safety separation standards, wide time windows, and constrained position shifting, with the aim of maximizing the total throughput. The work [5] aimed to reduce the waiting time in the runway holding point, by proposing a non-iterative real-time model, which can help air traffic controllers in decision-making in times of congestion on the ground at any airport. The proposed model shows that the number of sequenced aircrafts simultaneously directly influences the waiting times and makes pan. Moreover, factors such as landing time and operational restrictions influence the optimized sequencing as well as the waiting time of the aircraft on the ground. In particular, the separation minima possibly become the most important factor of influence when the SID is considered [6] in their work focuses on the real-time optimization of takeoff and landing operations at a busy terminal control area in case of traffic congestion. The formulations in their proposed mixed-integer linear programming investigate the trade-off between multiple performance indicators of practical objective while considering the safety constraints with a high precision. Paper [7] proposed a method that imputes departing flights into the aircraft sequence, sets up a dynamic model to consider time-varying variables, and constructs a specific genetic algorithm to solve the aircraft-sequencing problem. The authors in paper [8] put into test the ASP over multiple runways, under mixed-mode operations with the aim of reducing the total weighted tardiness of aircraft landings and departures at the same time. The ASP was modeled as a parallel machine scheduling problem with unequal ready times, target times, and deadlines. Greedy heuristics and metaheuristics were applied to obtain solutions in reasonable computation times.

Paper [9] presented the issue of integrated gate reallocation and taxiway scheduling, in which complex constraints related to runway restriction, gate assignment, and taxiway conflict are all added when determining the schedule. Contribution [10] shows the development of a multilevel optimization framework for the design and selection of departure trajectories, and the distribution of aircraft movements among these trajectories while taking the sequence and separation needs for aircraft on runways and along chosen trajectories in consideration. The authors in [11] present a robust optimization approach for metering aircraft departures under uncertainty in the taxi-out process. A mixed-integer linear programming model for runway sequencing and scheduling that incorporates uncertainty sets for the release time is proposed so as to determine in a dynamic way an ideal and robust sequence and schedule of aircraft taxi-out from the gate. In article [12], the authors show the development of a two-step optimization framework to select aircraft departure trajectories and the allocation of flights in these trajectories. The work [13] presents the first local search heuristic for the coupled runway sequencing (arrival and departure) and taxiway routing problems, based on the receding horizon (RH) scheme that consider the dynamic nature of the problem. Paper [14] presents a new multi-objective optimization model for the construction and assignment of optimal aircraft departure trajectories. In this problem—in addition to the two conventional objectives based on cumulative noise criteria and fuel burn—a new objective taking into account the flight frequency is considered. In addition, to take advantage of the combination of designing new routes and allocating flights to these routes, two different routes are considered simultaneously, and the distribution of flights over these two routes is addressed in parallel. The authors in paper [15] study optimal departure operations at airports in the context of departure metering. More precisely, they develop a stochastic dynamic programming framework for tactical management of pushback operations at gates and for determining the optimal number of aircraft to be directed to the runway queue from the metering areas.

4 Modelization

In order to optimize the problem, we considered four categories of aircrafts according to their performances. We also chose three standard departure routes which leading the north, east, and west noted, respectively, N, E, and W. After a study of cumulated data, we quantified the conflict between two aircrafts following two SIDs in a coefficient. We define C_i, $i - 1$ the coefficient which represents the conflict between the aircraft i and the preceding aircraft $i - 1$ as shown in Table 1.

We define:

- $D_{i,j}$: the delay in seconds that the aircraft i has to wait in order to take off after aircraft j.
- T_i: the time that the aircraft i needs to reach a certain altitude or flight level to liberate the SID (5000 ft in our model)
- Twy estimate: the time for aircraft i to reach the principal Taxiway
- Twy time: the time that the aircraft i spend on the principal taxiway.

We are going to compute the delay using the following algorithm:

Algorithm 1

```
X=[N,E,W]
(for i=1:3
sid(i )=Xᵢ
    (for J= 1:3
        c(i,j)= C Xᵢ,Xⱼ
     D i,j= c(i,j) × Cat i × 25
     end)
end)
```

We obtain the delay using the formula:

$$D_{i,j} = C_{i,j} \times Cat_i \times 25 \tag{1}$$

Then we replace: $j = i - 1$

$$D_{i,i-1} = C_{i,i-1} \times Cat_i \times 25 \tag{2}$$

Table 1 SID coefficient

i	j	$C_{i,j}$
W	N	3
W	E	1
W	W	4
E	N	3
E	E	4
E	W	1
N	N	4
N	E	3
N	W	3

After that, we are going to sequence the aircrafts in a descending order according to the estimated time of reaching the main taxiway so as to find the takeoff order following the first come first served concept FCFS: A = [Ai], with i ∈ {2; 3; : : : ; n}; is the order of takeoff of aircraft Ai according to FCFS, so the order is as follows:

{ A1, A2, : : : , An)}; therefore, Ai takes off at testf1 = tA1, as for A2, it can take off at

$$testf2 = max(testf1 + Di, i - 1, tA2) \qquad (3)$$

because an aircraft cannot take off before its estimated time of departure. This separation takes into account the regulatory separation. We define:

$$Ri = testfi - tAi \qquad (4)$$

the delay of the aircraft Ai with i is the takeoff order according to the FCFS concept.

> Concept of the scheduling algorithm:
> Once the n aircraft are activated, referring to the type of aircraft, it calculates the estimated taxiing time of each aircraft Ai, then following this time sequence the aircraft in a descending order to find the optimal permutation p.

After that we are going to sequence the aircrafts in a descending order from the estimated time of the holding point in order to find the takeoff order according to FCFS concept. A = [Aj], with j ∈ {p(1), p(2); p(3); : : ; p(n)}; is the takeoff order of the plane Aj according to FCFS, so the order is as follows:

{ Ap(1), Ap(2); Ap(3); : : : : ; Ap(n)}; therefore, A1 takes off at testfp1 = tAp(1), as for Ap(2) can take off at

$$testfp2 = max(testfp1 + Di, j , tAp(2)) \qquad (5)$$

We define:

$$Rpi = testfpi - tAp(i) \qquad (6)$$

the delay of the aircraft Ap(i) with i is the takeoff order according to the FCFS concept. The first aircraft A1 to take off will be among the most performing aircrafts in the selection.

TO preview the efficiency of our model, we have minimized the constraints assuming that all the aircrafts will follow the same SID or departure trajectory after takeoff.

Algorithm 2

> 1 Calculate ti for i=1,...,n
> 2 Find p
> 3 Calculate testfpi for i=1,...,n
> 4 Calculate the delay Rpi for i=1,...,n

5 Short Job First (SJF) Scheduling Concept

In this work, we will choose a more efficient scheduling method in order to:

- Ensure that each waiting aircraft will be handled in the shortest possible time.
- Minimize the waiting time.
- Use the airport infrastructure to the maximum.
- Balance the use of resources.
- Consider the priorities.
- Predict the traffic situation in the best way.

We will consider traffic sequencing as a system with non-primitive scheduling or without requisition. It will execute the tasks by choosing the shortest first (Short Job First—SJF) and give priority until it finishes the whole takeoff phase by freeing the SID or the departure trajectory for the following traffic.

This method is valid when we know the maximum execution time (the taken time by an aircraft in the SID or departure trajectory) in order to get a better average time of stay.

In our model, we will also assume that all the aircrafts will follow the same departure trajectory or SID to determine an exact execution time.

We will consider that the aircraft is the task of the process and the arrival time is the time to reach the waiting point, while the stay time will be the time between the waiting point and the moment when the departure trajectory or SID is released and the waiting time is the time that the planes have to wait before executing the flight.

6 Computational Study and Data

Table 2 shows the computational data following the FCFS algorithm with 'arrival time' is the estimated time for an aircraft to reach the holding point after the taxiing phase, 'job' is the execution time of the takeoff and crossing flight level 70 (7000 ft) phase and 'Execution end' is the estimated time to release the mentioned level.

After simulating a given situation containing a certain number of aircrafts in the first phase using the FCFS algorithm, a delay of 318.9 min was generated, knowing that each aircraft is misplaced in the sequence, which explains the generated accumulated delay during the taxiing phase, something that is unacceptable in the aviation sector because every wasted minute is worthy.

So, we decided to simulate the same situation using another sequencing method which is the SJF and we have obtained as a result 204.7 min as a generated accumulated

Table 2 FCFS concept application (22 aircrafts)

FIFO	Aircraft category	Arrival time (min)	Job	Execution	Execution end	Delay (min)
1	1	1.6	1.4	1.6	3	0
2	2	3.8	2.8	3.8	6.6	0
3	3	6	4.2	6.6	10.8	0.6
4	4	8.3	5.6	10.8	16.4	2.5
5	3	10	4.2	16.4	20.6	6.4
6	2	12	2.8	20.6	23.4	8.6
7	1	14.6	1.4	23.4	24.8	8.8
8	4	17	5.6	24.8	30.4	7.8
9	2	19	2.8	30.4	33.2	11.4
10	4	21	5.6	33.2	38.8	12.2
11	3	23	4.2	38.8	43	15.8
12	1	25	1.4	43	44.4	18
13	2	27	2.8	44.4	47.2	17.4
14	4	30	5.6	47.2	52.8	17.2
15	1	32	1.4	52.8	54.2	20.8
16	3	34	4.2	54.2	58.4	20.2
17	2	36	2.8	58.4	61.2	22.4
18	4	39	5.6	61.2	66.8	22.2
19	2	41	2.8	66.8	69.6	25.8
20	1	43	1.4	69.6	71	26.6
21	3	45	4.2	71	75.2	26
22	2	47	2.8	75.2	78	28.2
				Total delay		**318.9 min**

delay; a drop of 35% compared to the first model as detailed in Table 3.

The impact of the number of aircrafts

To show the impact of the number of the handled aircrafts in the same periods (peak hours), which is the main reason of this work, we have varied the number of traffic in situations ranging from 5 to 15 aircrafts in order of 5 as detailed in the Tables 4, 5, 6, 7, 8, and 9.

Five aircrafts computing data

See Tables 4 and 5.

Ten aircrafts computing data

See Tables 6 and 7.

Fifteen aircrafts computing data

See Tables 8 and 9.

Summary and graphic interpretation

We can see that the delay decreases significantly depending on the aircrafts number, which favors the use of SJF algorithm in the departure scheduling in comparison with the FCFS concept as detailed in Table 10 and illustrated in Fig. 2.

7 Conclusion

The objective of this paper is to develop an algorithm that optimizes the departure sequence by considering the aircraft categories, the spent time on the taxiway, on the runway, and during the climb in the departure trajectory. The used algorithm, based on the "Shortest Job First SJF" concept, sequenced the ground traffic according to the estimated time to reach the holding point in order to allow a better use of the departure trajectories. For simulation constraints, we have considered that all the aircrafts will follow the same trajectory after departure and grouped them in four main categories. We also have supposed that an optimized scheduling

Table 3 SJF concept application (22 aircrafts)

SJF	Aircraft Category	Arrival time (min)	Execution	execution end	Delay (min)
1	1	1.6	1.6	3.8	0
2	2	3.8	3.8	6.6	0
3	3	6	6.6	10.2	0.6
5	3	10	10.2	14.4	0.2
6	2	12	14.4	17.2	2.4
7	1	14.6	17.2	19	2.6
9	2	19	19	21.8	0
4	4	8.3	21.8	27.4	13.5
12	1	25	27.4	28.8	2.4
13	2	27	28.8	31.6	1.8
11	3	23	31.6	35.8	8.6
15	1	32	35.8	37.2	3.8
17	2	36	37.2	40	1.2
16	3	34	40	44.2	6
20	1	43	44.2	45.6	1.2
19	2	41	45.6	48.4	4.6
22	2	47	48.4	51.2	1.4
21	3	45	51.2	55.4	6.2
8	4	17	55.4	61	38.4
10	4	21	61	66.6	40
14	4	30	66.6	72.2	36.6
18	4	39	72.2		33.2
				Total delay	**204.7 min**

Table 4 FCFS concept application (5 aircrafts)

FIFO	CATEGORY	ARRIVAL TIME (MIN)	JOB	EXECUTION	EXECUTION END	DELAY (MIN)
1	1	1.6	1.4	1.6	3	0
2	2	3.8	2.8	3.8	6.6	0
3	3	6	4.2	6.6	10.8	0.6
4	4	8.3	5.6	10.8	16.4	2.5
5	3	10	4.2	16.4	20.6	6.4
						9.5

Table 5 SJF concept application (5 aircrafts)

SJF	CATEGORY	ARRIVAL TIME (MIN)	JOB	EXECUTION	EXECUTION END	DELAY (MIN)
1	1	1.6	1.4	1.6	3	0
2	2	3.8	2.8	3.8	6.6	0
3	3	6	4.2	6.6	10.8	0.6
5	3	10	4.2	10.8	15	0.8
4	4	8.3	5.6	15	20.6	6.7
						8.1

Table 6 FCFS concept application (10 aircrafts)

FIFO	CAT	ARRIVAL TIME (MIN)	JOB	EXECUTION	EXECUTION END	DELAY (MIN)
1	1	1.6	1.4	1.6	3	0
2	2	3.8	2.8	3.8	6.6	0
3	3	6	4.2	6.6	10.8	0.6
4	4	8.3	5.6	10.8	16.4	2.5
5	3	10	4.2	16.4	20.6	6.4
6	2	12	2.8	20.6	23.4	8.6
7	1	14.6	1.4	23.4	24.8	8.8
8	4	17	5.6	24.8	30.4	7.8
9	2	19	2.8	30.4	33.2	11.4
10	4	21	5.6	33.2	38.8	12.2
						58.3

Table 7 SJF concept application (10 aircrafts)

SJF	CATEGORY	ARRIVAL TIME (MIN)	JOB	EXECUTION	EXECUTION END	DELAY (MIN)
1	1	1.6	1.4	1.6	3	0
2	2	3.8	2.8	3.8	6.6	0
3	3	6	4.2	6.6	10.8	0.6
5	3	10	4.2	10.8	15	0.8
6	2	12	2.8	15	17.8	3
7	1	14.6	1.4	17.8	19.2	3.2
4	4	8.3	5.6	19.2	24.8	10.9
9	2	19	2.8	24.8	27.6	5.8
8	4	17	5.6	27.6	33.2	10.6
10	4	21	5.6	33.2	38.8	12.2
						47.1

Table 8 FCFS concept application (15 aircrafts)

FIFO	CATEGORY	ARRIVAL TIME (MIN)	JOB	EXECUTION	EXECUTION END	DELAY (MIN)
1	1	1.6	1.4	1.6	3	0
2	2	3.8	2.8	3.8	6.6	0
3	3	6	4.2	6.6	10.8	0.6
4	4	8.3	5.6	10.8	16.4	2.5
5	3	10	4.2	16.4	20.6	6.4
6	2	12	2.8	20.6	23.4	8.6
7	1	14.6	1.4	23.4	24.8	8.8
8	4	17	5.6	24.8	30.4	7.8
9	2	19	2.8	30.4	33.2	11.4
10	4	21	5.6	33.2	38.8	12.2
11	3	23	4.2	38.8	43	15.8
12	1	25	1.4	43	44.4	18
13	2	27	2.8	44.4	47.2	17.4
14	4	30	5.6	47.2	52.8	17.2
15	1	32	1.4	52.8	54.2	20.8
						147.5

Table 9 SJF concept application (15 aircrafts)

SJF	CATEGORY	ARRIVAL TIME (MIN)	JOB	EXECUTION	EXECUTION END	DELAY (MIN)
1	1	1.6	1.4	1.6	3	0
2	2	3.8	2.8	3.8	6.6	0
3	3	6	4.2	6.6	10.8	0.6
5	3	10	4.2	10.8	15	0.8
6	2	12	2.8	15	17.8	3
7	1	14.6	1.4	17.8	19.2	3.2
9	2	19	2.8	19.2	22	0.2
4	4	8.3	5.6	22	27.6	13.7
12	1	25	1.4	27.6	29	2.6
13	2	27	2.8	29	31.8	2
15	1	32	1.4	31.8	33.2	-0.2
11	3	23	4.2	33.2	37.4	10.2
8	4	17	5.6	37.4	43	20.4
10	4	21	5.6	43	48.6	22
14	4	30	5.6	48.6	54.2	18.6
						97.1

Table 10 FCFS and SJF summary delays

Numb of aircraft	5	10	15	22
FCFS delay	9.5	58.3	148	318.9
SJF delay	8.1	47.1	97.1	204.7

Fig. 2 FCFS and SJF delays

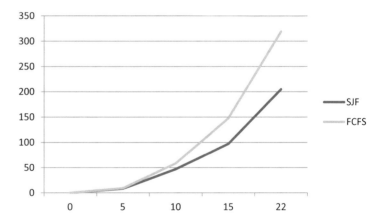

system already executed before the holding point. The achieved results are very satisfactory in comparison with the classical FCFS concept.

References

1. Annual Network Operations Report 2019. https://www.eurocontrol.int/publication/annual-network-operations-report-2019 (consulté le nov. 27, 2020).

2. ICAO, *Procedures for air navigation and air traffic management pans-atm doc 4444*. Place of publication not identified: ICAO, 2016

3. N. Schefers, J.J. Ramos González, P. Folch, J.L. Munoz-Gamarra, A constraint programming model with time uncertainty for cooperative flight departures. Transp. Res. Part C Emerg. Technol. **96**, 170–191 (2018). https://doi.org/10.1016/j.trc.2018.09.013

4. R. Prakash, R. Piplani, J. Desai, An optimal data-splitting algorithm for aircraft scheduling on a single runway to maximize throughput. Transp. Res. Part C Emerg. Technol. **95**, 570–581 (2018). https://doi.org/10.1016/j.trc.2018.07.031

5. H.F. Fernandes, C. Müller, Optimization of the waiting time and makespan in aircraft departures: A real time non-iterative

sequencing model. J. Air Transp. Manag. **79**, 101686 (2019). https://doi.org/10.1016/j.jairtraman.2019.101686

6. M. Samà, A. D'Ariano, P. D'Ariano, D. Pacciarelli, Scheduling models for optimal aircraft traffic control at busy airports: tardiness, priorities, equity and violations considerations. Omega **67**, 81–98 (2017). https://doi.org/10.1016/j.omega.2016.04.003

7. S. Caprì, M. Ignaccolo, Genetic algorithms for solving the aircraft-sequencing problem: the introduction of departures into the dynamic model. J. Air Transp. Manag. **10**(5), 345–351 (2004). https://doi.org/10.1016/j.jairtraman.2004.05.004

8. G. Hancerliogullari, G. Rabadi, A.H. Al-Salem, M. Kharbeche, Greedy algorithms and metaheuristics for a multiple runway combined arrival-departure aircraft sequencing problem. J. Air Transp. Manag. **32**, 39–48 (2013). https://doi.org/10.1016/j.jairtraman.2013.06.001

9. C. Yu, D. Zhang, H.Y.K. Henry Lau, A heuristic approach for solving an integrated gate reassignment and taxi scheduling problem. J. Air Transp. Manag. 62, 189–196 (2017). https://doi.org/10.1016/j.jairtraman.2017.04.006

10. V. Ho-Huu, S. Hartjes, J. A. Pérez-Castán, H. G. Visser, R. Curran, A multilevel optimization approach to route design and flight allocation taking aircraft sequence and separation constraints into account. Transp. Res. Part C Emerg. Technol. **117**, 102684 (2020). https://doi.org/10.1016/j.trc.2020.102684

11. M.C.R. Murça, A robust optimization approach for airport departure metering under uncertain taxi-out time predictions. Aerosp. Sci. Technol. **68**, 269–277 (2017). https://doi.org/10.1016/j.ast.2017.05.020

12. V. Ho-Huu, S. Hartjes, H.G. Visser, R. Curran, An optimization framework for route design and allocation of aircraft to multiple departure routes. Transp. Res. Part Transp. Environ. **76**, 273–288 (2019). https://doi.org/10.1016/j.trd.2019.10.003

13. U. Benlic, A.E.I. Brownlee, E.K. Burke, Heuristic search for the coupled runway sequencing and taxiway routing problem. Transp. Res. Part C Emerg. Technol. **71**, 333–355 (2016). https://doi.org/10.1016/j.trc.2016.08.004

14. V. Ho-Huu, S. Hartjes, H.G. Visser, R. Curran, Integrated design and allocation of optimal aircraft departure routes. Transp. Res. Part Transp. Environ. **63**, 689–705. https://doi.org/10.1016/j.trd.2018.07.006

15. H. Chen, S. Solak, Lower cost departures for airlines: Optimal policies under departure metering. Transp. Res. Part C Emerg. Technol. **111**, 531–546 (2020). https://doi.org/10.1016/j.trc.2019.12.023

A Multiagent and Machine Learning Based Denial of Service Intrusion Detection System for Drone Networks

Said Ouiazzane, Malika Addou, and Fatimazahra Barramou

Abstract

The objective of this research work was to propose a model based on Muli-Agent System and on machine learning techniques to detect Denial of Service (DoS) cyber-attacks targeting the networks of drones. The proposed model is autonomous, characterized by its high performance and enables the detection of known and unknown DoS attacks in UAV networks with high accuracy and low false-positives and false-negatives rates. This approach is intended to address the security vulnerabilities of drone-based infrastructures and to show how important this topic is, given that little attention is paid by the scientific community to the security aspect of drones. The detection of DoS attacks is an indispensable security measure to ensure the high availability of drone systems typically used in emergency situations (Intelligent Health and Public Safety Systems) where geospatial information is sensitive and highly critical. The proposed approach has made it possible to detect DoS attacks using multi-agent systems and the machine learning Decision Tree algorithm, which was chosen after testing several machine learning algorithms (Such as Random Forest, Decision Tree, Tree Ensemble, Naïve Bayes, Support Vector Machine…) on the CICIDS2017 which is a reference dataset used by researchers working on Network Intrusion Detection Systems (NIDS). The results of the experiment were conclusive and demonstrated the efficiency and effectiveness of our system in detecting DoS attacks with 100% of accuracy and with null rates of false positive and false negative.

Keywords

Denial of service • Geospatial data • Intrusion detection • Machine learning • Network of drones • CICIDS2017 • Muli-agent systems

1 Introduction

UAVs (Unmanned Aerial Vehicle) are widely used in different fields of expertise, and we have many examples of their use in the literature. Notably, the author of [1] used UAV technology to ensure multispectral imaging and in the field of agriculture to identify the difference in terrain. In another research work [2], the authors suggest using UAVs to embark an X-ray camera, IR, and metal detectors. This technology is also widely used in the field of e-commerce and delivery despite the limitations faced in terms of the strong impact of weight on the battery life of UAVs and the maximum distance that can be flown by the aircraft, knowing that operations to deliver small goods is already a reality that we live today, including several large technology companies (such as Amazon's Prime Air service) use UAVs for delivery of goods [3].

In the era of the COvid19 pandemic that the world is currently experiencing and that is constantly invading the entire planet, all nations have been mobilized to face the very rapid spread of the corona virus. In this sense, several countries have called upon UAV technology to significantly accelerate the control of this unprecedented spread. In particular, a flight of drones is being used in Australia to identify people with a "doubtful" respiratory profile [4]. China, the United Arab Emirates, Spain, and Kuwait have opted for UAV fleets (UAV Networks) to monitor and disinfect several cities to control the spread of Covid19 [5]. UAVs are also used to deliver medicines, food supplies, and other necessary goods weighing a few kilos. Currently, thousands of UAVs are deployed in every state of India to

S. Ouiazzane (✉) · M. Addou
ASYR RT, LaGeS Laboratory, Hassania School of Public Works, Casablanca, Morocco

F. Barramou
Geomatics Science Research Team (SGEO), LaGeS Laboratory, Hassania School of Public Works, Casablanca, Morocco

© The Author(s), under exclusive license to Springer Nature Switzerland AG 2022
F. Barramou et al. (eds.), *Geospatial Intelligence*, Advances in Science, Technology & Innovation,
https://doi.org/10.1007/978-3-030-80458-9_5

undertake similar actions, obviously with government authorization. These systems have demonstrated their effectiveness with a significant success rate in reducing the spread of the virus [5]. UAV systems can be equipped with communication tools and sent to suitable stations in the field to act as aerial mobile stations, to perform mobile relay station tasks to extend network coverage, or as end terminals [6–10]. In this case, UAVs are very effective in ensuring the high availability of telecom infrastructures in situations of natural disasters, war, and terrorism.

Given the very important use of UAVs in various fields of application including the control and the processing of emergency situations (Covid19, Public Safety…) where geospatial information is critical and urgent, UAVs are increasingly targeted by attackers and hackers to affect their security through the violation of geospatial data security policies, which is the key factor of success of missions carried out by UAVs. As a result, the security and protection of geospatial data transiting through UAV networks against cyber-attacks (that could impact the reliability, integrity, and availability of UAV control and application flows) are current issues that need to be taken seriously and studied in depth. As little attention has been paid to date by the scientific community to the detection of intrusions into UAV networks, and most research works focuses mainly on conventional problems and solutions to improve performance in terms of coverage, mobility, and reliability, while ignoring the security aspect. In this research work, we present a model of an intrusion detection system for UAV networks to protect geographic data against denial of service attacks, which are among the most common attacks against UAV infrastructures and fleets.

Today, cyber-attacks such as DoS (Denial of Service) and DDoS (Distributed Denial of Service) have become more notorious and widespread. DDoS attacks allow an attacker to exploit a large number of compromised machines (called zombies), which bombard one or more specific targets with seemingly legitimate requests. DDoS attacks are often effective, affordable, and do not require well forged technical expertise to instantiate them [11]. Throughout this paper, we will use the term DoS to refer to both DoS and DDoS cyber-attacks since they have the same operating principle and the difference lies in the maximum number of requests that can be sent.

DoS attacks can be instantiated by hackers and malicious persons (even by hacking amateurs who can use open source tools published on the Internet to launch such type of cyberattacks) to affect the availability of UAV infrastructures in order to prevent UAVs and their owners from carrying out their missions or to capture UAVs for illicit purposes. Denial of service attacks generally aim at depleting the resources of UAV systems to prevent the timely completion of important tasks [12]. Given the distribution, volume, transition speed, and particular qualities of geographic data [13], the use of classical DoS attack detection systems becomes inappropriate as the detection of this type of attack in typical UAV networks is a very complex task that needs to be simplified by opting for multi-agent technology, which is the most appropriate way to deal with the security of geographic data against DoS attacks.

2 Background of the Study

In this section, we will discuss some concepts and terminologies related to our research work. At the beginning, we will give an overview of UAVs, UAV fleets, and geospatial data circulating within UAV networks. Then we will move on to the security aspect of UAVs while identifying the various vulnerabilities and attacks to which UAV technology is exposed. Finally, we will conclude this section with some basic notions about Network Intrusion Systems (NIDS).

2.1 Drone Definition

A UAV is defined as an aircraft with no pilot on board. The UAV system is usually remotely controlled using a so-called Ground Control Station (GCS). The UAV is autonomous and can fly on its own following a pre-programmed flight plan, just as it can be remotely controlled using a smartphone or tablet [14].

UAVs are widely used in different fields of expertise, and we have many examples of their use in the literature. In particular, the author of [1] proposes to use UAV technology to take multispectral images and to identify the difference in terrain to enhance the agriculture productivity. In another research work [2], the authors suggest using UAVs to embark an X-ray camera, IR, and metal detectors. This technology is also widely used in the field of e-commerce and delivery despite the limitations faced in terms of the strong impact of weight on the battery life of UAVs and the maximum distance that can be flown by the aircraft, knowing that operations to deliver small goods is already a reality that we live today, including several large technology companies (such as Amazon's Prime Air service) which have already started to use UAVs for delivery of goods [3]. Finally, according to the author of [1], another use of the drone technology can be envisaged, which is the transport of blood and small medicines in Africa.

2.2 Fleet of Drones

A UAV fleet is a grouping of several UAVs that collaborate with each other to achieve more complex objectives [2]. In a UAV network, each UAV takes on its tasks to contribute to the achievement of the overall objective. We can consider four types of communication architectures that can be used in a UAV fleet: satellite communication architecture, centralized communication architecture, ad hoc communication architecture, and cellular communication architecture.

2.3 Geographic Data

Geospatial or geographical data refer to objects located on a given environment. According to the geometry and the role of data, we must distinguish between several types of data [15]:

- According to the geometry:
  - Vector data (point, line, and polygon)
  - Raster data (images: scanned IGN maps, aerial images)
  - Locatable data
- Depending on the role
  - Geographical references
  - Business data

Geographic data about the world are stored in thematic layers that can be linked together geographically. This concept, both simple and powerful, has proven its effectiveness in solving many concrete problems [16].

2.4 Security Aspect of a Fleet of Drones

In this section, we will focus on some of the vulnerabilities inherent to WIFI networks and those present as a result of the involvement of UAV technology in these networks. We will also look at some of the attacks to which UAVs are exposed and which can jeopardize the security of the application and control flows that pass through UAV networks.

Vulnerabilities of a UAV network. UAV networks are vulnerable to cyber-attacks due to the lack of a central entity responsible for monitoring suspicious activities that could target the network. Moreover, in an UAV network deployed under an ad hoc communication architecture, all nodes of the network participate in proposing routing table entries and do not assume the presence of malicious UAVs that may broadcast erroneous routing packets [2]. In addition, some vulnerabilities may occur that are inherent to the physical architecture of the UAVs [17], the communication media, the deployment architecture (cooperation and collaboration between the members of the fleet), and the possibility of inserting malicious UAVs into the network [18].

UAV networks are generally based on WIFI and use radio waves as their communication media. However, these radio links are characterized by their low bandwidth that can be easily saturated by a hacker's broadcast of false packets, so communication between the UAVs in the network can be disrupted [19].

The adhoc deployment architecture of a UAV fleet is not controlled due to its distribution and dynamic characteristics. In this type of communication architecture, the communication medium is opportunistic and shared between the UAVs in the network. As a result, the task of controlling the entry and the output of the nodes in the network becomes very delicate. In this case, a malicious drone could easily insert itself into the network and thus participate in proposing paths with optimal metrics that could falsify the entries in the routing table. Without forgetting that a malicious drone can always usurp the identity of the legitimate nodes to act on the routing table by broadcasting false packets or by replaying obsolete information (Example: Rushing attack) [20].

The mobility speed of UAVs is very high (speeds can reach 80 km/h or more), which continuously changes the network topology. Therefore, a routing protocol cannot distinguish a communication failure caused by UAV movements from the activities of a hacker trying to make the service unavailable [21].

Another very important vulnerability is present due to the very limited resources available to UAVs (CPU and RAM). Consequently, the drone resources can be easily consumed by attackers using, for example, sleep deprivation attacks [22] which consist in broadcasting a very large number of control messages to capture and bypass drone devices.

Attacks in an adhoc network. UAV fleets rely on wireless technology as a communication medium. Consequently, a UAV network can be targeted by different cyber-attacks, namely, eavesdropping and active interference [23]. This type of attack can be carried out by using a high-gain antenna within range of a UAV to listen, capture, or replay network traffic passing through the UAV fleet.

Given the importance and criticality of the application and control flows carried by a UAV network. Hackers are constantly aiming to circumvent these networks by, for example, eavesdropping on a network while placing an illegitimate UAV between two or more nodes in the fleet. They can also take advantage of the inherent vulnerabilities of UAV systems that may result from poor configuration of UAV networks, inadequate implementation, or protocol design problems [24]. Not to mention that the attacker can listen to the entire network by acting on the routing protocols to create the inputs and paths that allow him to redirect all traffic to his system.

2.5 Computer Security

Security refers to the ownership of a system, service, or entity. It is generally expressed by the principles of security [25], which are availability, integrity, and confidentiality [26]. According to [27] the principles of computer security are as follows:

- Confidentiality: This is about respecting the privacy of individuals and protecting their sensitive information from misuse or unauthorized disclosure.
- Integrity: Data cannot be altered or tampered without authorization. It also means that data stored in one part of a database system is consistent with other related data stored in another part of the database system (or on another system).
- Availability: The IT infrastructure (information, services, systems, equipment…) must function properly while remaining available at all times to serve authorized users.

2.6 Network Intrusion Detection System—NIDS

Intrusion Detection Systems (IDS) are among the most important defense tools against sophisticated and growing cyber-attacks targeting networks. Intrusion is defined as any computer activity aimed at illegitimately accessing information, eavesdropping on the network without authorization or completely destroying the system [28].

Intrusion detection is one of the most important aspects of cybersecurity [29]. Network Intrusion Detection Systems (NIDS) are indispensable devices for detecting violations of computer network security policies. A NIDS monitors and analyzes the traffic on an entire organization's network and informs administrators as soon as it detects suspicious activities that could have an impact on the proper functioning of the information system. There are two types of NIDS: The first type is the Signature-based NIDS (SNIDS); this type verifies the pattern matching against a database of signatures developed by security experts. The second type is the Anomaly Detection Based NIDS (ADNIDS); this one is based on the identification of a baseline profile of the network during its normal functioning and then triggers alarms as soon as a deviation from the reference profile is detected [30].

To study NIDS, several datasets are published on the Internet to be used by researchers working on the development and the improvement of intrusion detection systems. In particular, NSL-KDD and CICIDS2017 are reference datasets that are used in the development of NIDS, and these datasets generally classify network connections into five categories: Normal traffic, DoS network, probe, R2L, and U2R. Thus, network flows can be classified as follows [31]:

- DOS: Denial of Service is a category of attack in which the victim's resources are consumed until the system becomes unable to respond to legitimate requests.
- Probe: The purpose of this category of attack is to obtain information about the victim remotely via a port scan for example.
- U2R: Getting unauthorized access to local superuser privileges is a type of attack, where an attacker uses a normal account to log into a victim's system and attempts to gain root/admin privileges by exploiting certain vulnerabilities on the victim's machine, such as buffer overflow attacks.
- R2L: Having unauthorized access from a remote machine to manipulate the victim's system as needed.

3 State of the Art

In the literature, several research works have dealt with the problem of UAV security. In this section, we will highlight some studies conducted and approaches proposed by researchers to address the security aspect of UAVs.

3.1 Security Aspect in UAV Networks

Javaid et al. [32] examines and analyzes the privacy and security challenges posed by Covid19 contact monitoring systems. The author proposes some recommendations for automatic contact tracing systems and UAV-based surveillance systems. Thus, any implementation of such systems must take into account the security aspect in order to prevent the violation of the privacy of unsuspecting citizens against cybercrime.

In [33], UAVs are considered vulnerable to cyber-attacks due to the typology of their network and to their various physical components located in remote locations. Cyber-attacks can target UAV networks for causing control station malfunctioning, destruction and exfiltration of UAV data, denial of service, and for corruption and damage to information. The author of [34] presents a study of incidents that may be caused by UAVs deployed near airports while highlighting some research that has used sensor technologies to detect, to mitigate, and to prevent malicious UAVs. The author of [35] proposes Blockchain technology as a means to improve UAV security. He confirms that this technology is highly secure since it is based on private key cryptography and peer-to-peer networking. In this article [36], the author conducts an exhaustive study of the different UAV simulators available on the market as they are essential to address the security issues facing UAV technology such as fatal UAV failures, malicious attacks targeting this technology and disastrous abuses that have cost UAV users a lot of money. The author of the work [37] presents the different threats and vulnerabilities of Internet of Drones (IoD) while creating a taxonomy to categorize attacks according to vulnerabilities and threats related to the networking of UAVs in existing cellular infrastructures. In the research work [38], the author focuses on the different concerns of the European Commission mainly related to the processing of data transiting in UAV networks. This commission aims to ensure the protection of privacy and cybersecurity while opting for the identification and registration of all types of UAVs, their pilots, and operators. The authors of [39] provide an overview of the technologies used for the surveillance of UAVs and existing anti-drug systems. According to the authors, UAVs are increasingly used in many fields of application because of their ease of use and low cost of deployment, posing a very large threat to public safety and the privacy of citizens. To limit the impact of these threats, it is very important to deploy anti-drug systems in sensitive areas in order to detect, locate, and defend against intruding UAVs. According to the author of [40], UAVs rely largely on autonomous systems (autopilot), so it is necessary to develop a very robust autopilot system to counter any possible cyber-attack that could target UAVs. Thus, the author presents a biometric system ensuring the encryption of the communication flow between a UAV and a computerized control station. In [41], the drone is a technology that offers many opportunities to improve health care by opting for drone drug delivery to increase drug accessibility. Thus, the author describes the drone technology and highlights some regulatory considerations to be taken into account by the pharmacy and all stakeholders to protect the privacy of citizens from disclosure. The author of [42] focuses on potential cyber-attacks and provides some examples of cyber-attacks that have targeted drone technology. The author also provides an overview of the aviation cybersecurity framework to assess the level of maturity against global security standards. The author concludes his work by highlighting the measures that need to be taken seriously for the implementation of aviation cyber security in UAV systems. In [14], the author proposes a model of a network intrusion detection system for a fleet of UAVs. According to the author, the system is based on machine learning techniques to detect any deviation from the network baseline.

3.2 DoS Attacks Targeting a Drone

Studies on denial of service attacks that can target drones have been conducted by the scientific community. In particular, the author of [12] presents a framework to cope with DoS attacks targeting UAVs while offering some defense mechanisms to protect the UAV resources in terms of

memory, CPU, and network card. This approach consists in preventing the attacker from using all the available resources to the UAV. In [43], the author evaluates the effects of denial of service attacks that can target the two commonly used UAVs (3DR SOLO and AR.Drone 2.0). The author confirms that DoS attacks are triggered at the time of UAV navigation and states that experiments have shown the effectiveness of this type of attack. The author adds that DoS attacks can critically impact the proper functioning of UAV systems during navigation after bombarding the use of their resources by malicious persons and thus cause malfunctions in the UAV infrastructure, namely, Camera malfunctioning, loss of telemetry data, and loss or alteration of control command flows. The author also presented the security mechanism implemented on these UAVs to avoid availability problems after a certain distance has been exceeded. In addition, the authors of [44] performed intrusion tests on Parrot Bebop UAVs and demonstrated that these commercial UAVs are vulnerable to DoS attacks at the time of ARDiscovery connection, which could lead to an immediate shutdown of the UAV rotors in mid-flight if the vulnerability is exploited. The authors also found that this commercial brand of drone is vulnerable to Address Resolution Protocol (ARP) cache poisoning attacks, as exploitation of this vulnerability could disconnect the aircraft owner and thus lose the drone altogether. Finally, article [45] highlights three tools that can be used to launch denial of service (DoS) attacks while analyzing the behavior of the UAV (AR.Drone 2.0) at the time of these attacks. The author indicates that DoS attacks could impact the availability of the network, which leads to malfunctions in critical UAV applications (in particular the "video streaming" functionality).

3.3 Discussion

It is true that the scientific community is nevertheless beginning to be aware of the importance of the security aspect of UAVs used to carry out emergency missions such as smart health systems (Covid19 application, for example) and public safety systems. However, the work carried out within the state of the art section remains limited in terms of enhancing the security of UAV system components and does not address the intrusion detection mechanism in UAV networks to detect Denial of Service attacks. Most of the research work addressing the security aspect is limited to studying the vulnerabilities present in UAV-assisted networks and proposing some recommendations to be taken into consideration when using UAVs in civilian applications.

Several authors confirm that the involvement of UAVs adds further security challenges to UAV fleet networks. On the one hand, the network infrastructure is becoming increasingly complex, making it easier to carry out attacks that could jeopardize the security of UAVs. This makes the task of analysis and defense a very delicate one. On the other hand, highly sensitive and critical data transit and are stored in UAV systems. As a result, these systems can present risks related to the invasion of citizens' privacy in case of infiltration and violation of security policies. In addition, UAV platforms typically include a variety of modules to enable proper functioning, but this also makes them vulnerable to possible malicious attacks, i.e., the exploitation of UAV sensor inputs for malicious purposes. In some cases, UAVs are equipped with sensors to monitor environmental parameters in order to search for specific objects, and if an attacker succeeds in manipulating these parameters to misdirect the sensors, the information will be falsified and the mission will fail. In this case, public health and safety officers would be led to react erroneously to certain events tampered with by hackers or prevent them from identifying the objects they are looking for altogether. Given that the majority of UAVs are equipped with a Global Positioning System (GPS) module, and if an attacker manages to transmit erroneous GPS signals to the GPS module, the reliability and security of the UAV system will be compromised. Another potential vulnerability is in the communication module used by UAVs to exchange data and commands with their environment, including the control station. This communication module is generally based on Zigbee or Wi-Fi technologies. Although these technologies are very important for the proper functioning of UAVs, they also introduce another vulnerability in UAV systems, since adversaries can, at any time, intercept or alter wireless signals to block critical information or to prevent public safety and health officials from controlling UAVs outside a danger zone. Another possible threat is a denial of service attack where an attacker could keep a server unavailable after consuming all its resources by sending a large number of insignificant requests so that it receives no relevant data from the drones. In addition, the security mechanisms of UAV networks should at least ensure the high availability of UAV-based infrastructures and the geospatial data passing through them. However, the exchange of information between UAVs and between UAVs and the ground station requires low network latency. As a result, most of the security problems associated with UAV networks are generally due to the limited resources available to the UAVs and the time constraints imposed by the real-time nature of this type of networks.

To our knowledge and based on the research works studied in the state of the art, no work has addressed the detection of intrusions at the level of UAV networks. We believe that ensuring security in the detection of DoS-type cyber-attacks, which can target UAV-based infrastructures,

is currently a priority to ensure the smooth running of very urgent missions. Given the particularity of geospatial data that meet the challenges of Big Data in terms of high volume, high throughput, and the variety of application and control flows, our current research work is based on a DoS intrusion detection model to prevent the DoS attacks targeting the fleet of drones. This system can detect DoS cyber-attacks while relying on agent technology and on detection mechanisms based on machine learning techniques.

4 Proposed Approach

In this section, we will highlight our proposed model of an intrusion detection system to detect cyber-attacks and suspicious activities that could target the network of a fleet of drones. To do so, we will begin by providing an overview of our system to identify the location of our system in an UAV network. Then we will talk about our proposed model, its components, and its operating principle. Finally, we will show the detection mechanism (micro level) adopted by the system to detect known and unkown DoS cyber-attacks.

4.1 General View

In this research work, we propose an intrusion detection system for a fleet of UAVs used to treat and control emergency situations, similar to the intelligent health systems used to control the spread of the Covid19 pandemic and public safety systems.

Figure 1 shows one of the most recent uses of UAV technology. The diagram shows a fleet of UAVs that is used to capture health data from people in public places. The data collected by the various UAVs in the fleet are then sent to Big Data analysts to extract indicators to be displayed to health workers to monitor the spread of the pandemic. In this architecture (see Fig. 1), citizens' data are exposed to the outside world and can be subject to cyber-attacks that can violate the security and privacy of people observed without their knowledge.

To detect intrusions and other suspicious activities that could jeopardize the security of application and control flows, our NIDS system will be designed to capture all network traffic (Application and Control Flows). To do this, the system must be fitted with a high-gain antenna so that it can capture all the data circulating in the UAV network to send the captured traffic to the Intrusion Detection System (IDS) to detect any malicious activity or attack that could violate current security policies.

Fig. 1 Location of our system in a network of drones

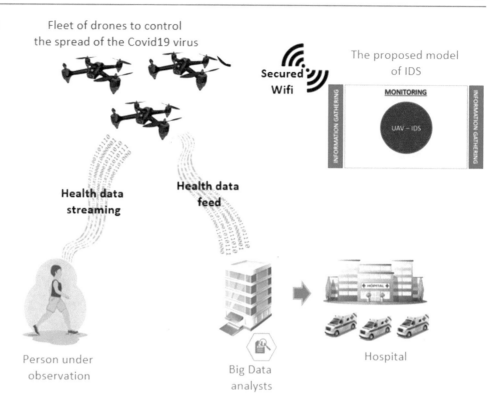

4.2 Inspirational Model

The intrusion detection model proposed in this research work was inspired by a model (see Fig. 2) that we have proposed in another previously published journal article [12].

As a reminder, the intrusion detection by the model [12] is provided by a set of seven agents and three modules, each of which is responsible for a specific task to collaborate together. The model is designed to detect known and unknown cyber-attacks based on agent technology and machine learning techniques.

4.3 Proposed Model

In this research work, we propose a DoS intrusion detection system for UAV networks. The proposed model is a second version of the previous model [12] which has been readapted to allow the detection of DoS and DDoS attacks that can target UAV networks (see Fig. 3).

Our model is mainly based on a set of three agents and two modules that interact and communicate with each other to carry out the task of detecting DoS cyber-attacks in UAV networks. The knowledge base is continuously fed using supervised and unsupervised machine learning techniques to ensure that all packets that have passed through the network are recognized and identified by the NIDS.

4.4 Components of the Proposed System

Given the complexity of UAV networks in terms of the mobility of fleet members and the rapid change in network topology. We opted for agent technology to simplify the problem, knowing that the multi-agent paradigm is the most appropriate solution to address the security issues in this type of dynamic networks. Thus, our system includes the following elements:

- Sniffer Agent (SA): This is the input of our system dealing with capturing the network traffic of the UAV fleet. This agent is equipped with a high gain antenna, and it has to be placed where it can cover all the members of the UAV fleet.
- Match Checker Agent (MCA): This is a reactive agent whose main task is to check the match of received packets against a knowledge base of known DoS attacks. If the network packet is recognized as a DoS attack by the knowledge base, the security administrator will be notified of the detected cyberattack and if the network packet is unknown yet, it will be handled using unsupervised machine learning techniques.
- Alert Manager Agent (AMA): This is an agent that takes care of the correlation of alerts to reduce the amount of false alarms and notifications. It also gives the administrator the ability to mark alarms and alerts as true or false alarms based on his expertise and investigations.

Fig. 2 Muliagent-based NIDS
for a fleet of drones (Redrawn
based on [12])

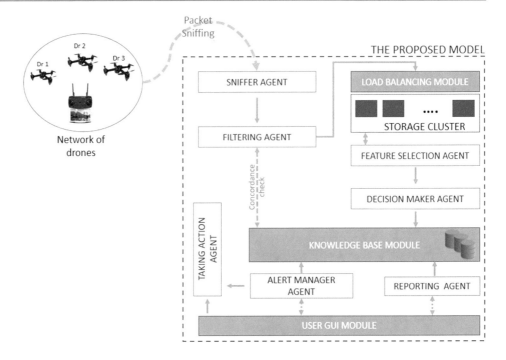

Fig. 3 The proposed model of
NIDS

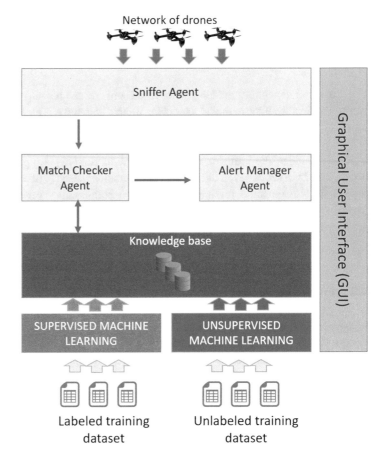

- Knowledge base module: This is a signature database containing the profiles of the DoS cyber-attacks and the baseline of normal operation. This database is enriched as new cyber-attacks are detected on the training of the system using supervised and unsupervised machine learning techniques.

4.5 Principle of Operation of the Proposed NIDS Model

The detection of known and unknown DoS intrusions is done in nine steps:

- Step 1: The system captures network packets based on a high gain antenna that can cover all members of the UAV fleet.
- Step 2: The captured packets are sent to the Match Checker Agent.
- Step 3: The Match Ckecker Agent (MCA) receives the network packets and checks their correspondence against the signature database of known attacks.
- Step 4: Depending on the result of the previous step, three scenarios can occur:
  - If the packet is suspicious: In this case the system will alert the security administrator to find out the details of the intrusion,
  - If the packet is normal: the packet will be ignored since it does not represent any security risk,
  - If the packet is unknown: it will be automatically sent to the unsupervised machine learning module.
- Step 5: If the packet is unknown, it will be handled using unsupervised machine learning techniques.
- Step 6: The detection result generated by unsupervised machine learning will be stored in the knowledge base in the form of a signature so that from now, such network packets will be recognized in step 3.

4.6 Intrusion Detection Mechanism (Micro Level of Detection)

The detection mechanism (see Fig. 4) adopted by our NIDS system is generally based on machine learning techniques to detect the set of known and unknown cyber-attacks. To do so, the system will be trained on an annotated training dataset containing all the characteristics of the known attack packets.

The detection of known attacks will be ensured by supervised machine learning techniques so that the system can detect existing known intrusions into the dataset. Unknown intrusions are detected using unsupervised machine leaning techniques to classify the packets according to similarities (clusters). The resulting clusters are processed by the voting method to annotate the data and then the system have to be re-trained on the new dataset to recognize the packets. In this present work we will just focus on the first phase based on supervised machine learning to detect known DoS attacks and for the second phase for detecting unkown attacks it will be covered in a future work.

5 Tests and Results

In this part, we will try to apply several supervised machine learning algorithms in order to choose the most adequate technique for the detection of DoS cyber-attacks whose behaviors are already known. In this research work, we will

Fig. 4 DoS attack detection mechanism (Micro level)

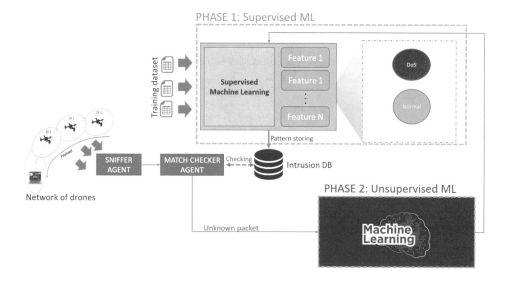

just address the detection mechanism to train our NIDS model on known DoS attacks. This experimentation allows us to enrich the knowledge base with signatures and patterns of known DoS attacks. However, the detection mechanism for unknown DoS attacks will be the subject of another article in the near future.

5.1 Training Dataset

To develop the detection mechanism of our system, the CICIDS2017 Dataset will be used to train our system on the known DoS attacks. This dataset contains benign traffic and most recent cyber-attacks. CICIDS2017 resembles actual network traffic (PCAP) and includes data annotated by analyzing network traffic using CICFlowMeter [46].

We will use the CICIDS2017 dataset which contains all known DoS and DDoS attack traces. Since UAV networks are based on TCP/IP protocols, the network packets that transit through UAV networks are similar to those transiting in traditional computer networks. Moreover, in a network packet, we just focus on the headers and TCP/IP characteristics, which is generic and can be applied to UAV networks. In addition, we do not yet have a public dataset captured from UAV networks, so we can use the CIDIDS2017 containing all traces of the most recent DoS attacks. As a result, training our system on this dataset will allow us to recognize all known DoS attacks that could target a UAV network, knowing that DoS attacks are the most widespread and easiest to instantiate against UAV networks.

Since the dataset used is light despite the fact that it contains millions of lines, the learning time is very short and does not require to opt for feature reduction. We considered all the attributes to be relevant while letting the Machine Learning based classification techniques do their job without having to reduce the number of attributes.

5.2 Machine Learning Techniques

Random Forest. The use of Random Forest (see Fig. 5) as a machine learning technique to detect DoS attacks gave good results with 99.98% accuracy.

The Random Forest algorithm allowed the detection of DoS intrusions with an error rate of $1.33.10^{-4}$, and Table 1 shows the confusion matrix of the algorithm.

Decision Tree. The Decision Tree algorithm has given very good results in the detection of DoS attacks where the accuracy reaches 100%. The workflow used is shown in Fig. 6.

The confusion matrix of the algorithm (see Table 2) shows a very good precision with a zero error rate.

Tree Ensemble. The Learning Tree Ensemble machine technique has given good results with an accuracy of 99.99%. Figure 7 highlights the followed work flow to have a good detection of DoS attacks with a very low error rate $(6.64.10-5)$.

Table 3 gives the confusion matrix of the algorithm and shows the number of erroneous results that do not exceed 3 DoS attacks detected as normal (False Negative).

Naive Bayes. The Naive Bayes algorithm gave poor results with a poor accuracy of 68.7%. Figure 8 illustrates the used workflow to apply this machine learning technique with an error rate as high as 0.31.

The confusion matrix (see Table 4) shows that the algorithm detected 14,126 normal and non-intrusive packets as DoS attacks (False Positive).

SVM—Support Vector Machine. Figure 9 presents the workflow adopted to test the detection of DoS cyber-attacks using the SVM (Support Vector Machine) algorithm. This technique allowed good intrusion detection with an accuracy of 99.99% and an error rate of 2.21.10–5.

The confusion matrix below (see Table 5) shows that the algorithm was able to effectively detect DoS attacks with a

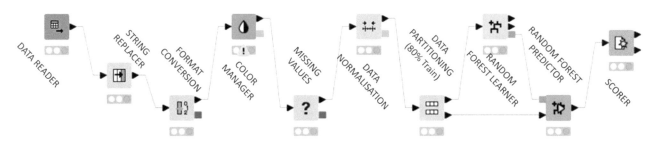

Fig. 5 Random Forest Workflow

Table 1 Confusion matrix of the Random Forest Algorithm

	Benign	DDos
Benign	19,558	1
DDos	5	25,585

single false negative (considering a DoS attack as a normal packet).

PNN—Probabilistic Neural Network. The PNN (Probabilistic Neural Network) algorithm was evaluated following the workflow in Fig. 10.

The PNN machine learning technique has given conclusive results in the detection of DoS cyber-attacks. The algorithm was able to achieve an accuracy of 99.99% with an error rate of $4.43.10^{-5}$. The confusion matrix below (see Table 6) shows that the algorithm was able to ensure a very

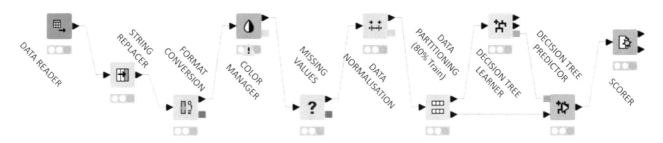

Fig. 6 Decision Tree Workflow

Table 2 Confusion matrix of the Decision Tree Algorithm

	Benign	DDos
Benign	19,526	0
DDos	0	25,623

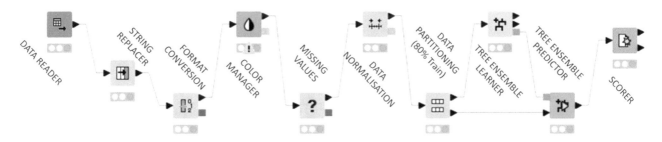

Fig. 7 Tree Ensemble Workflow

Table 3 Confusion matrix of the Tree Ensemble Algorithm

	Benign	DDos
Benign	19,567	0
DDos	3	25,579

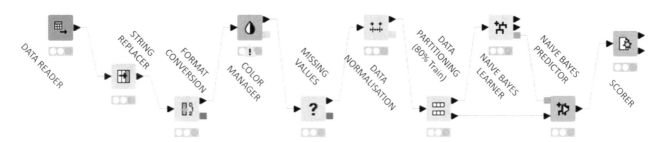

Fig. 8 Naïve Bayes Workflow

Table 4 Confusion matrix of the Naïve Bayes Algorithm

	Benign	DDos
Benign	5294	14,126
DDos	0	25,729

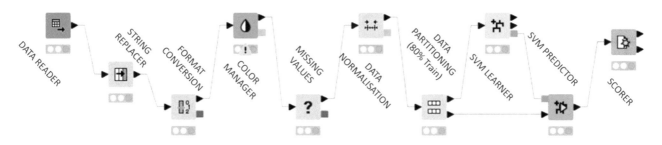

Fig. 9 SVM Workflow

Table 5 Confusion matrix of the SVM Algorithm

	Benign	DDos
Benign	19,567	0
DDos	1	25,581

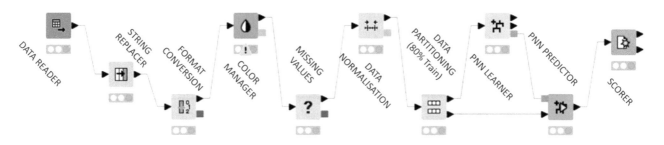

Fig. 10 PNN Workflow

Table 6 Confusion matrix of the PNN Algorithm

	Benign	DDos
Benign	19,655	0
DDos	2	25,492

efficient detection of DoS cyber-attacks with 2 false negatives.

Gradient Boosted Trees. Figure 11 below shows the workflow of the Gradient Boosted Trees technique.

The Gradient Boosted Trees algorithm gave good detection results with an accuracy of 99.99% and an error rate of $6.64.10^{-5}$. The confusion matrix (see Table 7) shows that

this technique was effective in detecting DoS attacks except that it gave 3 false negatives.

K Nearest Neighbor. The K Nearest Neighbor algorithm (see Fig. 12) has also been tested and has given satisfactory results in the detection of DoS cyber-attacks.

The confusion matrix in Javaid et al., illustrates effective detection with 2 results that are false negatives.

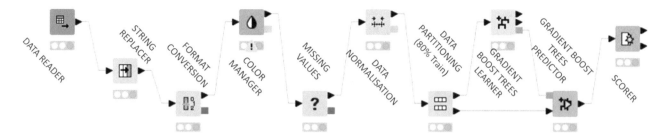

Fig. 11 Gradient Boosted Trees Workflow

Table 7 Confusion matrix of the Gradient Boosted Trees Algorithm

	Benign	DDos
Benign	19,560	0
DDos	3	25,586

Table 8 Confusion matrix of the K Nearest Neighbor Algorithm

	Benign	DDos
Benign	19,572	0
DDos	2	25,575

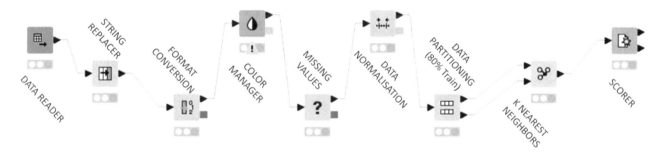

Fig. 12 K Nearest Neighbor Workflow

5.3 Recap of the Achieved Results

In this section, we have elaborated a consolidated table summarizing all the obtained results after using different machine learning techniques to train our supervised model to detect DoS attacks that can target a network of UAVs. Table 9 summarizes the detection results using the different algorithms.

According to the analysis of the obtained results, most techniques are very effective in detecting DoS attacks with an accuracy between 99.98% and 99.99%. Among the tested algorithms, we have the Decision Tree technique which gave very satisfactory results with 100% in terms of accuracy and zero false positive and false negative rates. The generated model using this algorithm proves to be the most appropriate to implement in our DoS intrusion detection system in order

to effectively detect DoS cyber-attacks that could jeopardize the availability of UAV-based infrastructures. Furthermore, the Naive Bayes algorithm cannot be used in this case due to its low accuracy, which does not exceed 68.7%.

To conclude this part, we will opt for the Decision Tree algorithm to train our system on known DoS attacks in order to feed the "Knowledge Base" module with all known attack behaviors. For the unknown attack detection mechanism, it will be treated very soon in another scientific work.

6 Conclusion and Perspectives

In this work, we have proposed a model of DoS and DDoS intrusion detection system in UAV networks. The proposed model is based on agent technology and machine learning

Table 9 Recap table of the used machine learning algorithms

Technique	False	Correct	Error	Accuracy (%)
Random Forest	6	45,143	1.33×10^{-4}	99.98
Decision tree	0	45,149	0.0	100
Tree ensemble	3	45,146	6.64×10^{-5}	99.99
Naive bayes	14,126	31,023	0.31	68.7
SVM	1	45,148	2.21×10^{-5}	99.99
Gradient boosted trees	3	45,146	6.64×10^{-5}	99.99
PNN	2	45,147	4.43×10^{-5}	99.99
K-nearest neighbor	2	45,147	4.43×10^{-5}	99.99

techniques. In addition, experimentation and analysis of supervised machine learning techniques have been carried out to select the most suitable algorithm with the best possible accuracy (Decision Tree with 100% of accuracy). The proposed model is very effective in detecting known and unknown DoS attacks that can target UAV networks. The system is autonomous and has a knowledge base that is fed as new DoS attacks are detected. Besides, the use of multi-agent technology makes the system modular and distributed, which eliminates the SPoF (Single Point of Failure) points and simplifies troubleshooting in case of an incident at the level of the NIDS (replacing the damaged component instead of replacing the entire system). Finally, the fact of using distributed storage ensures real-time network packet processing without causing network latency.

References

1. C. Lum, M. Mackenzie, C. Shaw-Feather, E. Luker, M. Dunbabin, Multispectral imaging and elevation mapping from an unmanned aerial system for precision agriculture applications, in *Proceedings of the 13th International Conference on Precision Agriculture*, St. Louis, MO, USA, 31 July–4 August 2016
2. M. Hamza, A. Jehangir, T. Ahmad, A. Sohail, M. Naeem, Design of surveillance drone with X-ray camera, IR camera and metal detector, in *Proceedings of the 2017 Ninth International Conference on Ubiquitous and Future Networks (ICUFN)*, Milan, Italy, 4–7 July 2017, pp. 111–114
3. Amazon. Amazon Prime Air (2016), https://www.amazon.com/Amazon-Prime-Air/b?node=8037720011. Accessed 11 April 2018
4. Draganfly selected to globally integrate breakthrough health diagnosis technology immediately onto autonomous camera's and specialized drones to combat coronavirus (covid19) and future health emergencies. https://apnews.com/Globe%20Newswire/dc01344350423d7d64c99ebbe8fb7548
5. Drone technology: A new ally in the fight against COVID-19, https://www.mdlinx.com/internal-medicine/article/6767. Accessed 15 April 2020
6. X. Li et al., The public safety wireless broadband network with airdropped sensors, in *Proceedings of the 2015 IEEE China Summit and International Conference on Signal and Information Processing (ChinaSIP)*, July 201, pp. 443–475
7. Z. Fadlullah et al., A dynamic trajectory control algorithm for improving the communication throughput and delay in UAV-aided

networks, in *IEEE Network*, vol. 30, no. 1, Jan–Feb. 2016, pp. 100–105
8. X. Li et al., Drone-Assisted Public Safety Wireless Broadband Network, in *Proceedings of the 2015 IEEE Wireless Communications and Networking Conference Workshop (WCNCW)*, Mar. 2015, pp. 323–28
9. A. Merwaday, I. Guvenc, UAV assisted heterogeneous networks for public safety communications. *Proceedings of the 2015 IEEE Wireless Communications and Networking Conference Workshop (WCNCW)*, Mar. 2015, pp. 329–34. [6]
10. G. Baldini et al., Survey of wireless communication technologies for public safety, in *IEEE Communications Surveys & Tutorials*, vol. 16, no. 2, May 2014, pp. 619–41
11. P. Gasti, G. Tsudik, E. Uzun, L. Zhang, DoS and DDoS in named data networking, in *2013 22nd International Conference on Computer Communication and Networks (ICCCN)*, Nassau, 2013, pp. 1–7, https://doi.org/10.1109/ICCCN.2013.6614127.
12. J. Chen, Z. Feng, J. Wen, B. Liu, L. Sha, A container-based DoS attack-resilient control framework for real-time UAV systems, in *2019 Design, Automation & Test in Europe Conference & Exhibition (DATE)*, Florence, Italy, 2019, pp. 1222–1227. https://doi.org/10.23919/DATE.2019.8714888
13. M.R. Luaces, N.R. Brisaboa, J.R. Paramá, J.R. Viqueira, A generic architecture for geographic information systems
14. S. Ouiazzane, F. Barramou, M. Addou, Towards a multi-agent based network intrusion detection system for a fleet of drones. Int. J. Adv. Comput. Sci. Appl. (IJACSA), 11(10) (2020). https://doi.org/10.14569/IJACSA.2020.0111044
15. Zipline (2017), http://www.flyzipline.com. Accessed 11 April 2018
16. J.-A. Maxa, Architecture de communication sécurisée d'une flotte de drones (2017)
17. S. Ouiazzane, M. Addou, F. Barramou, Toward a network intrusion detection system for geographic data (2020)
18. SCRIBD Website, https://fr.scribd.com/document/483184992/sig-esri-pdf. A. Kim, B. Wampler, J. Goppert, I. Hwang, H. Aldridge, Cyber attack vulnerabilities analysis for unmanned aerial vehicles. Infotech@ Aerospace, (2012)
19. A.Y. Javaid, W. Sun, V. K. Devabhaktuni, M. Alam, Cyber security threat analysis and modeling of an unmanned aerial vehicle system, in *Homeland Security (HST), 2012 IEEE Conference on Technologies for* (IEEE, 2012), pp. 585–590
20. P. Yi, Z. Dai, S. Zhang, Y. Zhong, A new routing attack in mobile ad hoc networks. Int. J. Inf. Technol. **11**(2), 83–94 (2005)
21. Y.-C. Hu, A. Perrig, D. B. Johnson, Rushing attacks and defense in wireless ad hoc network routing protocols, in *Proceedings of the 2nd ACM workshop on Wireless security*, (ACM, 2003), pp. 30–40
22. J. Elston, E. W Frew, D. Lawrence, P. Gray, B. Argrow, Net-centric communication and control for a heterogeneous unmanned aircraft system. J. Intell. Robotic Syst. 56(1–2), 199–232 (2009)

23. M. Pirretti, S. Zhu, N. Vijaykrishnan, P. McDaniel, M. Kandemir, R. Brooks, The sleep deprivation attack in sensor networks: analysis and methods of defense. Int. J. Distrib. Sens. Netw. **2**(3), 267–287 (2006)

24. B. Kannhavong, H. Nakayama, Y. Nemoto, N. Kato, A. Jamalipour, A survey of routing attacks in mobile ad hoc networks. IEEE Wirel. Commun. **14**(5), 85–91 (2007)

25. H. Debar, M. Dacier, A. Wespi, Towards a taxonomy of intrusion-detection systems. Comput. Netw. **31**(8), 805–822 (1999)

26. S. Ouiazzane et al., ICSSD '19 2019—A multi-agent model for network intrusion detection

27. Ghernaouti-Hélie et al., Sécurité informatique et réseaux cours avec plus (2013), http://medias.dunod.com/document/9782100599127/Feuilletage.pdf

28. D. K. Sharma, K.S. Nikhil, An approach for collaborative decision in distributed intrusion detection system (2016), http://citeseerx.ist.psu.edu/viewdoc/download?doi=10.1.1.735.1337&rep=rep1&type=pdf

29. Rathore et al, Real time intrusion detection system for ultra- high speed environments (2016). https://doi.org/10.1007/s11227-015-1615-5

30. I. Obeidat, N. Hamadneh, M. Alkasassbeh, M. Almseidin, M.I. AlZubi, Intensive pre-processing of KDD Cup 99 for network intrusion classification using machine learning techniques (2019)

31. Javaid et al., A deep learning approach for network intrusion detection (2016). https://doi.org/10.4108/eai.3-12-2015.2262516

32. L. Dhanabal, S.P. Shantharajah, A study on NSL-KDD dataset for intrusion detection system based on classification algorithms. Int. J. Adv. Res. Comput. Commun. Eng. (2015)

33. M. R. Hussein, A. B. Shams, E. H. Apu, K. A. Al Mamun, M. S. Rahman, Digital surveillance systems for tracing COVID-19: privacy and security challenges with recommendations (2020)

34. B. Siddappaji, K.B. Akhilesh, Role of cyber security in drone technology, in *Smart Technologies*, ed. by K. Akhilesh, D. Möller (Springer, Singapore, 2020). https://doi.org/10.1007/978-981-13-7139-4_13

35. G. Lykou, D. Moustakas, D. Gritzalis, Defending airports from UAS: a survey on cyber-attacks and counter-drone sensing technologies. Sensors **20**, 3537 (2020)

36. T. Rana, A. Shankar, M. K. Sultan, R. Patan and B. Balusamy, An intelligent approach for UAV and drone privacy security using Blockchain methodology, in *2019 9th International Conference on Cloud Computing, Data Science & Engineering (Confluence)* (Noida, India, 2019), pp. 162–167. https://doi.org/10.1109/CONFLUENCE.2019.8776613

37. A. Mairaj, A. I. Baba, A. Y. Javaid, Application specific drone simulators: recent advances and challenges. Simul. Modell. Pract. Theory **94**, 100–117 (2019), ISSN 1569-190X. https://doi.org/10.1016/j.simpat.2019.01.004

38. Gaurav Choudhary, Vishal Sharma, Takshi Gupta, Jiyoon Kim, Ilsun You, Internet of Drones (IoD): Threats, Vulnerability, and Security Perspectives – 2018

39. E. Bassi, "European Drones Regulation: Today's Legal Challenges," 2019 International Conference on Unmanned Aircraft Systems (ICUAS), Atlanta, GA, USA, 2019, pp. 443–450, doi: https://doi.org/10.1109/ICUAS.2019.8798173.

40. X. Shi, C. Yang, W. Xie, C. Liang, Z. Shi, J. Chen, Anti-drone system with multiple surveillance technologies: architecture, implementation, and challenges. IEEE Commun. Mag. **56**(4), 68–74 (2018). https://doi.org/10.1109/MCOM.2018.1700430

41. A. Singandhupe, H.M. La, D. Feil-Seifer, Reliable security algorithm for drones using individual characteristics from an EEG signal. IEEE Access **6**, 22976–22986 (2018). https://doi.org/10.1109/ACCESS.2018.2827362

42. Connie A. Lin, Pharm.D., Karishma Shah, M.S., Lt Col Cherie Mauntel, B.S.Pharm., Sachin A. Shah, Pharm.D., Drone delivery of medications: Review of the landscape and legal considerations, American Journal of Health-System Pharmacy, Volume 75, Issue 3, 1 February 2018, Pages 153–158, https://doi.org/10.2146/ajhp170196

43. M. Pyzynski, "Cybersecurity of the Unmanned Aircraft System (UAS)," 2020 International Conference on Unmanned Aircraft Systems (ICUAS), Athens, Greece, 2020, pp. 1265–1269, doi: https://doi.org/10.1109/ICUAS48674.2020.9213922.

44. Gabriel Vasconcelos , Rodrigo S. Miani , Vitor C. Guizilini and Jefferson R. Souza, Evaluation of DoS attacks on Commercial Wi-Fi-Based UAVs, 2019 (International Journal of Communication Networks and Information Security)

45. M. Hooper et al., Securing commercial WiFi-based UAVs from common security attacks, in *MILCOM 2016—2016 IEEE Military Communications Conference* (Baltimore, MD, 2016), pp. 1213–1218. https://doi.org/10.1109/MILCOM.2016.7795496

46. G. Vasconcelos, G. Carrijo, R. Miani, J. Souza, V. Guizilini, The impact of DoS attacks on the AR. Drone 2.0, in *2016 XIII Latin American Robotics Symposium and IV Brazilian Robotics Symposium (LARS/SBR)*, (Recife, 2016), pp. 127–132. https://doi.org/10.1109/LARS-SBR.2016.28

47. Online content, https://www.unb.ca/cic/datasets/ids-2017.html

48. D.C. He, S. Guizani, Mohsen drone-assisted public safety networks: the security aspect

Toward a Deep Learning Approach for Automatic Semantic Segmentation of 3D Lidar Point Clouds in Urban Areas

Zouhair Ballouch, Rafika Hajji, and Mohamed Ettarid

Abstract

Semantic segmentation of Lidar data using Deep Learning (DL) is a fundamental step for a deep and rigorous understanding of large-scale urban areas. Indeed, the increasing development of Lidar technology in terms of accuracy and spatial resolution offers a best opportunity for delivering a reliable semantic segmentation in large-scale urban environments. Significant progress has been reported in this direction. However, the literature lacks a deep comparison of the existing methods and algorithms in terms of strengths and weakness. The aim of the present paper is therefore to propose an objective review about these methods by highlighting their strengths and limitations. We then propose a new approach based on the combination of Lidar data and other sources in conjunction with a Deep Learning technique whose objective is to automatically extract semantic information from airborne Lidar point clouds by enhancing both accuracy and semantic precision compared to the existing methods. We finally present the first results of our approach.

Keywords

Lidar • Deep learning • Semantic segmentation • Urban environment

Z. Ballouch (✉) · R. Hajji · M. Ettarid
College of Geomatic Sciences and Surveying Engineering, IAV Hassan II, Rabat, Morocco
e-mail: z.ballouch@iav.ac.ma

R. Hajji
e-mail: r.hajji@iav.ac.ma

M. Ettarid
e-mail: m.ettarid@iav.ac.ma

1 Introduction

Several challenges are facing contemporary cities such as urban sprawl, environmental degradation, climate change, etc. Understanding these issues and predicting their impact can only be achieved through a deep and rigorous analysis of the urban environment. In this context, 3D city models are today positioned as powerful tools to address several needs about urban planning and sustainable development. Monitoring of the dynamics of cities, urban space management, construction design, and environmental studies are some appealing examples where a 3D city model is needed [1]. To respond to several city challenges, 3D city models are intended to be semantically rich to meet the requirements of urban planning and monitoring. Currently, Lidar techniques are recognized as powerful tools for producing 3D city models by offering very accurate and dense 3D point clouds at a large scale.

Semantic segmentation is an essential step to automatically design a rich 3D city model from Lidar data. It consists of assigning a semantic label for each group of point clouds (or a group of pixels in the case of images) based on homogeneous criteria [2] (Fig. 1).

The segmentation of 3D Lidar point clouds has been widely investigated in the literature leading to several notable achievements. However, this is still an active research trend until the challenges about geometric and semantic accuracy as well as robustness and performance of the proposed methods are to be resolved.

Currently, there is a lot of interest in developing Deep Learning (DL) techniques for analyzing 3D spatial data. Thanks to their potential for processing huge amounts of data corresponding to large scale and complex urban areas with good performance in terms of accuracy and efficiency, DL methods revolutionizes the field of computer vision and are the state-of-the-art in object detection and semantic segmentation [4, 5].

According to the literature, several developments have been conducted in the field of segmentation of 3D Lidar

© The Author(s), under exclusive license to Springer Nature Switzerland AG 2022
F. Barramou et al. (eds.), *Geospatial Intelligence*, Advances in Science, Technology & Innovation,
https://doi.org/10.1007/978-3-030-80458-9_6

Fig. 1 3D semantic representation [3]

point clouds. The developed methods can be classified into three families. The first one is based on the raw point cloud, the second one is based on a product derived from the cloud, mainly Digital Surface Model (DSM), while the third one combines original point clouds and other data sources (aerial image, land map, etc.) [1].

Several research teams have stated that the combination of Lidar data with other sources (aerial image, satellite image, etc.) is promising, thanks to the altimeter accuracy of the 3D point clouds and the planimetric continuity of the images [6]. This motivated us to conduct our research in this field where we propose to design a methodology based on the integration of Lidar data and other sources with the aim to enhance the quality of the semantic segmentation results for urban scenes.

In the next sections, we propose to give a global overview about the main developments in semantic segmentation by highlighting the strengths and the weakness of the developed approaches. Section 2 gives an overview about the main developed methods for automatic segmentation of Lidar point clouds. Then, Sect. 3 presents some DL approaches for semantic segmentation. The discussion of the main outcomes is the subject of Sect. 4. While Sect. 5 proposes the basic guidelines as well as the preliminary results of a new approach based on our investigations and the outcomes of the literature review. Finally, the paper ends with a conclusion.

2   Automatic Segmentation of 3D Point Clouds

Point cloud segmentation is an essential step for various applications. Besides clarifying the spatial relationships between point clouds and facilitating pattern recognition, the segmentation improves the quality of subsequent classifications. This process partitions a cloud of points into a set of segments characterized by spatial and/or geometric coherence. The definition of this coherence forms the critical part of the segmentation process [7]. Numerous segmentation approaches have been developed and applied to 3D Lidar data. In this section, we mainly focus on the general research methods that are widely used for the segmentation of 3D point clouds. Three families of approaches exist to perform a semantic segmentation of Lidar point clouds. The first one is based on the raw point cloud (Direct approaches). The second one is based on a product derived from the cloud (Derived Product Based Approaches). While the third one combines original point clouds and other data sources (aerial image, land map, etc.) (Hybrid approaches) [6].

2.1   Direct Approaches

Direct approaches are applied to 3D raw point clouds without any sampling method. Among the benefits of this family of approaches, we can cite the preservation of the original characteristics of data, including accuracy and topographical relationships. On the other hand, we raise some shortcomings and gaps that hinder the effectiveness and the relevance of this family of approaches, mainly the need of a too high computing time and a rather large memory.

In the literature, many studies have been based on direct approaches. Among the developed methods, Lee [8] has proposed a segmentation process based on 3D surface detection, specifically by using Lidar raw data directly without any prior interpolation. This method allows automatic division of the point cloud into two classes: ground and buildings, considered as the main objects of an urban scene. The method of [9] proposes a cluster analysis of 3D airborne Lidar data by using a slope adaptive neighborhood system based on accuracy, point density, and distance between 3D point clouds in order to define the neighborhood between the measured points. According proximity and local continuity, points that are on the same surface are connected [9]. The method gives good results in extracting vertical walls and modeling objects with a precision of few centimeters. Lari [10] proposed a method for segmentation of planar patches using Lidar data. In this approach, the authors used an adaptive cylinder for establishing the neighborhood of each point by considering surface trend and density. This definition of neighborhood positively influences the calculation of segmentation attributes (vegetation, flat and gable roofs, walls …etc.). The approach demonstrates efficiency and reliability for both airborne Lidar and Mobile Mapping Systems data. Finally, a segmentation method applied to a mobile and airborne mapping system has been proposed by [11] where the main objective is to bypass the drawbacks of point-based classification techniques; its principle is based on grouping point clouds in regions with similar characteristics. The proposed methodology demonstrates a high potential in classification of both terrestrial and airborne Lidar data.

2.2 Derived Product Based Approaches

Since direct approaches require a very high processing time and large storage capacity, many researchers recommend the transformation of 3D data into 2D in order to have a regular form that is easy to manipulate. This is the principle of derived based approaches which are based on derived products from Lidar data such as DSM (Digital Surface Model) and 2D images. This family of approaches offers a wide range of advantages such as the ease of handling and the efficiency of data processing. However, these approaches require a 2D transformation of 3D data or voxels representations which result in a huge loss of geometric and radiometric information, and thus a loss of precision due to the resampling operation.

In the literature, there is a large number of approaches that have been developed for the segmentation of 3D point clouds from the regular data generated from the point clouds. Among these approaches, Yuan [12] proposed a new technique called "Pointseg" that allows a real time semantic segmentation of road objects based on spherical images where the structure of the proposed network is based on SqueezeNet [13] and SqueezeSeg [14]. The proposed network has three main functional layers: (1) fire layer, (2) squeeze reweighting layer, and (3) enlargement layer. The results show compatibility with robot applications by

achieving competitive accuracy with 90 frames per second on a single GPU (Graphics Processing Unit) and high efficiency when tested with KITTI 3D objet detection dataset. Milioto [15] proposed a semantic segmentation approach called RangeNet++. This approach has been applied to Lidar data recorded by a rotating Lidar sensor in order to enable the autonomous vehicles to make the best decisions in a timely manner. The authors proposed a projection based 2D CNN (Convolutional Neural Network) processing of Lidar data and used a range image representation of each laser scan to perform the semantic inference. The results show that this method outperforms the state of the art both in runtime and accuracy. Moreover, a new approach for semantic labeling of unstructured 3D point clouds has been proposed in [16]. The authors proposed a framework that applies CNN on multiple 2D image views of the Lidar data based on two steps: (1) generation of two types of images: depth composite view and RGB view and (2) labeling each pair of bidimensional image views by means of CNNs. After that, they project back the semantized images. This approach showed good results when evaluated using a dataset called Semantic-8. Another method for Lidar data segmentation using voxel structure and graph-based clustering was proposed by [17]. The authors used a geometric method that not require any radiometric information. The process consists of three steps: (1) voxelisation of 3D point clouds, (2) calculation of geometric cues, and (3) the graph-based clustering. The method has demonstrated good results mainly for complex environment and non-planar areas, compared to several segmentation methods proposed in the literature. Riegler and Osman Ulusoy [18] proposed a method called "OctNet" as a novel tridimensional representation for point clouds labeling, which enables 3D CNN that are both high resolution and deep. The method was evaluated using Rue-Mong2014 dataset [19] and achieved good results. Finally, another work has been proposed by [20] where the authors evaluated various bidimensional image models using four datasets which are DUT1, NC, DUT2, and KAIST. The results, compared to those of direct approaches, show that the use of bidimensional image models give an interesting improvement in computational efficiency with a little loss of precision. Furthermore, the authors concluded that 2D image models are better suited to real-time segmentation of outdoor areas.

2.3 Hybrid Approaches

Despite the simplicity and the efficiency of Derived Product Based Approaches, several researchers argued that Lidar data need to be combined with other data sources (aerial photos, satellite images, etc.) to take benefits from the planimetric continuity of images and the altimetric precision of 3D point clouds [6]. Several investigations in this field have shown promising results in terms of accuracy and quality of the segmentation. However, despite their performance, these approaches have many disadvantages related to memory requirements, difficulty of handling and implementation, and the need to have a minimum difference in the time of acquisition of the two types of data.

The first method has been proposed by [21] for automatic building detection from 3D point clouds and multispectral imagery. This method is capable of detecting different urban objects (industrial buildings, urban residential, etc.) of different shapes with very high precision. The authors of [22] applied a multi-filter CNN for semantic segmentation based on the combination of 3D point clouds and high-resolution optical images, and then they used a MRS (Multi-Resolution Ssegmentation) for delimiting the contours of objects. The results show that this approach improves the overall accuracy over other methods using Potsdam and Guangzhou datasets and is more suitable for the processing of objects with a regular shape such as cars and buildings. Furthermore, Xiu [23] proposed a new method to study the influence of integrating two types of data which are aerial images and 3D point clouds for semantic segmentation which shows an accuracy of 88%. Additionally, a new semantic segmentation study combining images and 3D point clouds has been proposed by [24] by adopting the DVLSHR (Deeplab-Vgg16 based Large-Scale and High-Resolution) model which is satisfactory for semantic segmentation of large-scale scenes when compared to other methods developed in the literature using CityScapes dataset. Another approach called SPLATNet was proposed in [25]. This approach has been tested with RueMonge2014 dataset [19] where an Intersection Over Union score was computed for all classes in order to evaluate the semantic segmentation results. The proposed approach scores well among the state-of-the-art algorithms for semantic segmentation. Recently, [26] proposed a new methodology for semantic segmentation which grasps bidimensional textural appearance and tridimensional structural characteristics in an integrated framework. The authors evaluated this approach using ScanNet Dataset [27]. The method has demonstrated good results compared to 3DMV (3D-Multi-View) and SplatNet (Sparse lattice Networks) approaches. Similarly, Li [28] designed a 3D real-time semantic map using 3D point clouds and images of road scenes. The method consists of using a CNN to segment 2D images acquired by a camera, and then the semantic segmentation results and the 3D point clouds are fused to generate a unified point cloud with an associated semantic information. The proposed technique is effective for several complex tasks including autonomous driving, robot navigation, etc.

2.4 Summary

3D Lidar data segmentation methods can be grouped into: Direct approaches, Derived Product Based Approaches, and Hybrid approaches. The direct approaches are the least used in the literature because they require a very large storage capacity and are very demanding in processing and computing time. Despite their limitations, their strengths lie in the preservation of the characteristics and the original topological relationships of the point cloud. Derived Product Based Approaches are the most dominant, simplest, and quickest approaches in the literature. However, the resampling operation applied to the point cloud causes a huge loss of information and so a loss of precision of the segmentation process. Finally, approaches combining 3D Lidar data and other sources allow improving the accuracy of the segmentation. However, these approaches do not accept large time differences between the acquisition of Lidar and images and require a very high storage capacity and a very important processing time (Table 1).

Actually, the development of DL methods offers a best opportunity to satisfy the need of computer vision field and demonstrates a high potential in semantic segmentation in terms of accuracy and efficiency. Their performance in segmentation process would enhance the quality of the results. The next section tries to give a brief overview of researches addressing DL in semantic segmentation.

3 Contribution of DL to Semantic Segmentation

Actually, DL methods revolutionize the field of computer vision and demonstrate good performance in semantic segmentation by solving a wide range of difficult problems in this field [29]. In this section, we examine some DL techniques used in semantic segmentation of Lidar data acquired in urban areas.

PointNet is a reference network which opened the way for the use of DL techniques for semantic segmentation of Lidar data [30]. Its performance, combined with its ease of implementation, makes it a perfect baseline for semantic segmentation of 3D point clouds. The core principle of PointNet is to implement the permutation invariance of the points in a cloud directly into the network. To evaluate its performance, the authors used the Stanford 3D dataset where data are annotated with 13 classes (floor, chair, table, etc.). PointNet has demonstrated satisfactory results compared to the literature. Similarly, Qi [31] proposed a hierarchical DL model called "PointNet++" in order to process a set of points that have been sampled in metric space in a hierarchical

Table 1 Advantages and disadvantages of the different segmentation approaches

Approach	Advantages	Disadvantages
Direct approaches	– Preserve the original topological relationships of point cloud	– Expensive – Few developed programs
Derived product based approaches	– Easy and fast drive – Requires few parameters	– Loss of information and accuracy due to re-sampling – False data caused by resampling step – Errors accumulation
Hybrid approaches	– Accurate – Efficient	– Expensive – Require a minimum difference in time of acquisition of the two types of data

manner. To test this approach, four datasets have been used, namely, ModelNet40, MNIST, SHREC15, and ScanNet. The results show that the proposed approach is more suitable to process point sets robustly and efficiently compared to other existing methods. Besides, this methodology introduced hierarchical feature learning and captures spatial features at different scales which is important in case of objects of different sizes. Another semantic segmentation approach named SegCloud was proposed in [32]. The proposed approach combines the advantages of trilinear interpolation, neural networks, and FC-CRF (Fully Connected Conditional Random Fields). The authors used the trilinear interpolation to transform voxels predictions to raw 3D points, then the FC-CRF allows overall consistency, and fine semantic segmentation. The authors evaluated the performance of the proposed algorithm using four multi-scale datasets about indoor or outdoor scenes (NYU V2, S3DIS, KITTI, and Semantic3D). The results show that CRF allows a significant improvement of the network and a high ability to extract the contours of objects in a very clear way. Moreover, a novel fully CNN approach for semantic segmentation of images named SegNet has been developed by [33]. It consists of an encoder-decoder structure based on the convolution layers of the VGG-16 algorithm. The architecture of SegNet is symmetrical and allows precise positioning of abstract features with good spatial locations. CamVid dataset has been used to evaluate the performance of the proposed method. This dataset is divided into two sets: the first contains 367 images used for training the model while the second contains 233 images used for performance evaluation. The results show that this algorithm gives good results and achieves very high scores in the case of semantic segmentation of road environments. Furthermore, Landrieu and Simonovsky [34] proposed a new Lidar approach applicable for large 3D Lidar data where the main objective is to divide the point clouds into simple forms. The process is based on three main steps: (1) a new concept called a superpoint graph to encode the relationships between object parts by edge attributes is proposed, (2) a neural network is used for the representation of each simple shape, and (3) two public datasets (S3DIS and Semantic3D) are used to

improve the average of mIOU (mean Intersection Over Union). In addition, Qi [35] proposed a 3D object detection approach based on collaboration between Haugh Voting and point set network called VoteNet. It is a geometric method that does not require any radiometric information but shows clear improvements over hybrid methods. Additionally, Yang [36] proposed a new large-scale urban semantic segmentation framework by integrating multiple aggregation levels (point-segment-object) of features and contextual features for road facilities recognition from 3D Lidar data. This study achieved very satisfactory results with an object recognition accuracy of more than 90%. Finally, Hu [37] developed a new neural network architecture called "RandLA-Net" that directly uses 3D Lidar data based on point sampling in a random manner. In order to reduce the point density, to avoid loss of information caused by the resampling step, the authors proposed a new local feature aggregation module. Compared to the literature, the proposed approach demonstrates a good performance in terms of precision, calculation time and is not demanding a fairly large memory.

4 Discussion

Today, 3D city models allow better understanding of urban spaces which is crucial for optimal management of cities. They are capable to meet several needs related to simulation and decision making processes. However, most of 3D city models lack rich semantics about urban knowledge and are far to respond to several challenges about smart and sustainable cities.

In computer vision, semantic segmentation is defined as the assignment of a class to each coherent region of an image [2] or 3D point clouds. Many recent studies have shown the effectiveness of DL in this context [30, 34–37]. The first experiments of approaches dedicated to semantic segmentation of 3D point clouds began by the use of conventional image processing programs by transforming the 3D Lidar data into regular shapes (for example, series of images) as in the case of the approach proposed by [16] that requires a transformation of 3D point clouds to 2D images. Other DL techniques are based on the transformation of the Lidar data into a grid of voxels that have a regular form as the case of the SegCloud method that was proposed by [32]. These regular representations do not really allow a clear writing of the particular organization of Lidar data which limits the performance of this type of approach [34]. Besides, the voxel representation does not take into account the small details of 3D forms.

Several research teams have proposed a range of dedicated approaches directly analyzing Lidar data. Among these approaches, the PointNet approach, proposed by [30], operates at the point level, which allows a very fine segmentation. This method is adapted to 3D point clouds acquired in indoor scenes, but it requires a necessary adaptation or additional training to be adapted to large datasets [32]. Similarly, the PointNet++ method is applied to the raw point clouds [25] without any sampling operation, which saves the initial information [35]. This method has demonstrated better performance in semantic segmentation and object classification [35]. However, it shows some limitations, namely, large computation and memory cost [38, 39]. Furthermore, this approach is not able to aggregate the scene context around the object centers due to more clutter and inclusion of neighboring elements [35], and also lacks a relevant specification of the spatial connectivity between points [25]. We note that "PointNet" and "PointNet++" have not been tested on data acquired by a large scale airborne mapping system that contains more complicated urban geographic features [23]. Recently, several approaches have been developed for processing of large-scale 3D point clouds. In this context, we find the SPG method that allows the preprocessing of 3D Lidar data as super-graphs in order to subsequently apply a neural network to assign a semantic label for each group of points [15]. The main advantage of this approach is its ability to handle large point clouds simultaneously by cutting point clouds into simple shapes that are easier to classify than points, but despite the low number of network parameters, this approach is high demanding in terms of time of processing required by super-graph construction and geometric partitioning [15]. We can state that most of the existing semantic segmentation approaches require a variety of blocks partitioning steps, pre/post–processing as well as the construction of graphs. In contrary, the "RandLA-Net" approach is able to directly process large scale 3D Lidar data in a single pass with high efficiency (1 million points in a single pass) without any pre-processing or post-processing steps compared to the existing methods [39].

Finally, semantic segmentation is an active research trend which aims to reach robust methods to extract semantics from dense point clouds or images. The construction of these models from Lidar data requires designing new approaches capable of extracting the maximum amount of semantic information about a large-scale urban environment with high accuracy and efficiency. Our research tries to respond to this challenge by proposing an innovative hybrid approach which aims to enhance the quality of semantic segmentation of airborne Lidar point clouds.

5 Our Approach

The literature review about DL techniques that address semantic segmentation of Lidar point clouds shows that this is clearly a field that requires further research in order to improve the accuracy and the performance of the segmentation process. This has motivated us to conduct research in this field in order to propose an innovative approach for semantic segmentation of airborne Lidar data based on a hybrid solution. In this section, we expose the first guidelines and preliminary results of our proposed research in this context.

5.1 Methodology

We propose to design a DL approach based on the combination of 3D airborne Lidar data and aerial images for semantic segmentation of airborne Lidar point clouds corresponding to large-scale urban environments. Our methodology is expected to give better results in terms of precision and robustness to recognize 3D objects of urban scenes and associate them a rich semantic. Figure 2 summarizes the general workflow of our approach.

Our approach relies on the combination of the geometry of Lidar data and the spectral information of images. It is based on the use of raw data in order to retain the original characteristics and topological relationships of 3D point clouds. The first step consists of applying semantic segmentation to drone images which results will be integrated with Lidar data in order to refine the quality of the segmentation process (part 2). The test of the performance and the reliability of the proposed approach will be performed through several large-scale datasets. In the next section, we present and analyze the preliminary results related to the first step of the wokflow (Part1).

5.2 Preliminary Segmentation

Semantic segmentation from drone images is a first step of the general workflow. The results will be then integrated with Lidar point clouds to enhance the segmentation process. High spatial–resolution of data acquired by drones makes it possible to discriminate the different urban objects and associate them a semantic label. In this context, several DL

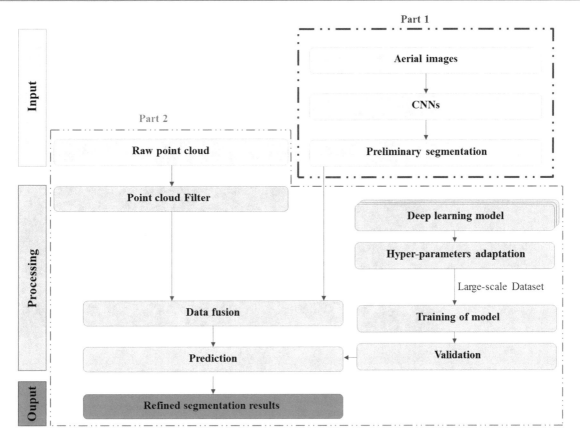

Fig. 2 The general workflow of our approach

techniques applied to drone images have been proposed in the literature [40–43]. To our knowledge, there is no literature review about the evaluation of the existing techniques. This is why we had to conduct several tests to evaluate different models (Unet, Vgg_Unet, Resnet50_Unet, Segnet, Vgg_Segnet, and Resnet50_Segnet) in terms of precision and calculation time in order to choose the most suitable one for semantic segmentation of drone images.

5.2.1 Data

The case study consists of 400 large-scale drone images with a high resolution of 6000 * 4000 px (24Mpx) and an altitude of 5–30 m above ground which are available for free download (https://dronedataset.icg.tugraz.at). The images are annotated with 20 classes: tree, gras, other vegetation, dirt, gravel, rocks, water, paved area, pool, person, dog, car, bicycle, roof, wall, fence, fence-pole, window, door, and obstacle. Some examples of the dataset are shown in Fig. 3.

Another data is used for the evaluation of the process. It is relative to an urban zone of the city of Nador (Morocco), where the images was acquired with a ground resolution at 100 m flight height of 3.5 cm and resolution of 12 MegaPixel.

5.2.2 Results

For the implementation of the DL models used in this study, we used the Keras library and Google Colaboratory as a cloud computing server. Google Colaboratory is a free Google tool that allows performing computational simulations with support of Python and some other libraries. For conducting the tests, 80% of the dataset is used for training the model while 20% serves as testing data. In this section, we present the results about the evaluation of both accuracy and time of calculation of the segmentation process applied to the selected models.

Accuracy assessment

The semantic segmentation realized according the tested models is evaluated through two parameters: (1) accuracy and (2) frequency weighted IU (f.w.IU). Accuracy metric is the ratio of the number of correct predictions to the total number of input samples. While the frequency weighted IU defines the variations on region intersection over union (IU) used in target detection [44]. These metrics are obtained using the equations below:

Fig. 3 Examples of classified
Drone images from the dataset

$$Accuracy = \frac{Number\ of\ correct\ predictions}{Total\ number\ of\ predictions\ made}$$

$$\text{f.w. IU} = \Sigma i\left(\Sigma j U i j\ \Sigma k \Sigma j U i j\right)\frac{U i i}{\Sigma j U i j + \Sigma j U j i - U i i}\ [44]$$

where k represents the number of classes. The symbol u_{ij} corresponds to the number of samples belonging to category i in ground truth and are classified in class j in segmentation results [44].

The evaluation results according accuracy and frequency weighted IU are reported in Table 2 and Table 3 respectively.

Even though we conducted the tests with a limited number of epochs, we reached good results for both accuracy and frequency weighted IU with the different models. According the results, we can say that the "Resnet50_Unet" model outperforms the others both in accuracy and frequency weighted IU metrics.

Training duration

Besides the accuracy of the segmentation, we also evaluated the efficiency of the tested models in terms of processing time. The results are reported in Table 4.

According to the statistics in Table 4, we can state that the processing time is relatively negligible and all models require almost the same computation time with a bit difference of the Vgg-Unet model which requires slightly more time than the other ones.

The preliminary tests were necessary to test the performance of the selected models. According the results, the "Resnet50_Unet" has been elected as the most suitable model for semantic segmentation of drone images to be adopted in our approach. This model has been applied to the case study about the urban area in Morocco. The corresponding semantic segmentation results are shown in Fig. 4 and validated by comparison to the field reality.

6 Summary

In the previous section, we presented the first results of the general workflow of our approach. It consists of semantic segmentation of drone images as a first step of the process. The general objective is to integrate the preliminary results of the image segmentation process with Lidar data in order to enhance the quality of the segmentation in terms of accuracy and performance. We performed a series of experiments to compare the capabilities of the different DL techniques for semantic segmentation of urban objects using Drone images.

Table 2 Comparison of accuracy between the DL models

		Unet	Vgg_Unet	Resnet50_Unet	Segnet	Vgg_Segnet	Resnet50_Segnet
	Accuracy	0.71	0.76	0.85	0.72	0.7215	0.82

Table 3 Comparison of frequency-weighted IU between the DL models

	Unet	Vgg_unet	Resnet50_Unet	Segnet	Vgg_segnet	Resnet50_Segnet
Frequency_Weighted_IU	0.56	0.63	0.76	0.58	0.56	0.72

Table 4 The required time for the segmentation process

	Unet	Vgg-Unet	Resnet50-Unet	Segnet	Vgg-Segnet	Resnet50_Segnet
Epochs	1310 s	1403 s	1225 s	1217 s	1208 s	1281 s
Epoch 1	1269 s	1385 s	1202 s	1198 s	1160 s	1222 s
Epoch 2	1243 s	1366 s	1219 s	1178 s	1159 s	1175 s
Epoch 3	1248 s	1319 s	1229 s	1154 s	1161 s	1171 s
Epoch 4	1205 s	1287 s	1209 s	1152 s	1163 s	1172 s
Total time (s)	6275	6760	6084	5899	5851	6021
Total time (m)	105	113	101	98	97	100

Fig. 4 Examples of semantic segmentation results

The results show that that all tested models give good results in terms of accuracy and frequency weighted IU. However, the Resnet50_Unet model scores well in both parameters. Hence, it has been selected as the most suitable one for semantic segmentation of drone images among the others. We should note that the quality of the results can be further improved by using a powerful dataset with more training data and by augmenting the number of epochs.

Finally, for a better evaluation of the performance of different DL models, we propose to use other types of datasets, as well as to apply the models to other images acquired in other different urban contexts.

7 Conclusion

In this paper, we have proposed a literature review about semantic segmentation methods of 3D Lidar point clouds based on DL. Several DL models have been presented and analyzed by highlighting their advantages and their limitations. We then presented the first guidelines about our proposed methodology which aims at developing a DL approach based on integrating 3D Lidar point clouds and aerial images for semantic segmentation in a large-scale urban environment. We aim to improve the object recognition accuracy and the efficiency of the existing methods.

As a first step of our approach, we investigated the performance of some DL models in terms of accuracy and performance for semantic segmentation of drone images by conducting several tests. In the next steps, our method will be tested on several datasets to confirm the reliability and the performance of the proposed approach.

References

1. A. Bellakaout, Extraction automatique des batiments, végétation et voirie à partir des données Lidar 3D. Thèse de docteur de l'institut agronomique et vétérinaire Hassan II, Maroc (2016)
2. L. Haifeng, Unsupervised scene adaptation for semantic segmentation of urban mobile laser scanning point clouds. ISPRS J. Photogramm. Remote. Sens. **169**, 253–267 (2020)
3. B. Kim, Highway driving dataset for semantic video segmentation. School of Electrical Engineering Korea Advanced Institute of Science and Technology (KAIST), South Korea (2016)
4. J. Castillo-Navarro, Réseaux de neurones semi-supervisés pour la segmentation sémantique en télédétection. Colloque GRETSI sur le Traitement du Signal et des Images, Lille, France. hal-02343961 (2019)
5. A. Garcia-Garcia, A review on deep learning techniques applied to semantic segmentation. arXiv:1704.06857v1 [cs.CV] (2017)
6. M. Awrangjeb, Automatic detection of residential buildings using LIDAR data and multispectral imagery. ISPRS J. Photogram. Remote. Sens. **65**, 457–467 (2010)
7. J. Ravaglia, Segmentation de nuages de points par octrees et analyse en composantes principales. GTMG 2014, Mar 2014, Lyon, France. hal-01376473 (2014)
8. I. Lee, Perceptual organization of 3D surface points, photogrammetric computer vision. ISPRS Comm. III. Graz, Austria. **XXXIV** part 3A/B. ISSN 1682-1750 (2002)
9. S. Filin, Segmentation of airborne laser scanning data using a slope adaptive neighborhood. ISPRS J. Photogramm. Remote. Sens. **60**, 71–80 (2006). https://doi.org/10.1016/j.isprsjprs.2005.10.005 (2005)
10. Z. Lari, An adaptive approach for segmentation of 3D laser point cloud, in *ISPRS Workshop Laser Scanning*, Calgary, Canada (2011)
11. Z. Lari, A. Habib, Segmentation-based classification of laser scanning data, in *ASPRS 2012 Annual Conference Sacramento*, California, 19–23 Mar 2012
12. W. Yuan, PointSeg: real-time semantic segmentation based on 3D LiDAR point cloud. arXiv:1807.06288v8 [cs.CV] (2018)
13. F.N. Iandola, Squeezenet: Alexnet-level accuracy with 50x fewer parameters and <1mb model size. CoRR abs/1602.07360 (2016)
14. B. Wu, Squeezeseg: convolutional neural nets with recurrent CRF for real-time road-object segmentation from 3d lidar point cloud. CoRR abs/1710.07368 (2017)
15. A. Milioto, RangeNet++: fast and accurate LiDAR semantic segmentation. German Research Foundation under Germany's Excellence Strategy, EXC-2070 - 390732324 (PhenoRob) as well as grant number BE 5996/1–1, and by NVIDIA Corporation (2019)
16. A. Boulch, SnapNet: 3D point cloud semantic labeling with 2D deep segmentation networks. Comput. Graph. (2017)
17. Y. Xu, Voxel- and graph-based point cloud segmentation of 3d scenes using perceptual grouping laws. ISPRS Ann. Photogramm. Remote. Sens. Spat. Inf. Sci. **IV-1/W1** (2017)
18. G. Riegler, A. Osman Ulusoy, Octnet: learning deep 3d representations at high resolutions, in *Proceedings of the IEEE Conference on Computer Vision and Pattern Recognition*, vol. 3 (2017)
19. H. Riemenschneider, A. Bódis-Szomorú, Learning where to classify in multi-view semantic segmentation, in *Proceedings of the European Conference on Computer Vision (ECCV)* (2014)
20. Y. Liu, Comparison of 2D image models in segmentation performance for 3D laser point clouds. Neurocomputing (2017)
21. M. Awrangjeb, Automatic detection of residential buildings using LIDAR data and multispectral imagery. ISPRS J. Photogramm. Remote Sens. **65**, 457–467 (2010)
22. Y. Sun, Developing a multi-filter convolutional neural network for semantic segmentation using high-resolution aerial imagery and LiDAR data. ISPRS J. Photogramm. Remote Sens. (2018)
23. H. Xiu, 3D semantic segmentation for high-resolution aerial survey derived point clouds using deep learning (Demonstration), in *Information Systems (SIGSPATIAL'18)*, 6–9 Nov 2018, Seattle, WA, USA, ed. by F. Banaei-Kashani, E. Hoel (ACM, New York, NY, USA, 2018)
24. R. Zhanga, Fusion of images and point clouds for the semantic segmentation of large scale 3D scenes based on deep learning. ISPRS J. Photogramm. Remote Sens. (2018)
25. H. Su, V. Jampani, Splatnet: sparse lattice networks for point cloud processing, in *Proceedings of the IEEE Conference on Computer Vision and Pattern Recognition* (2018), pp. 2530–2539
26. H.-Y. Chiang, A unified point-based framework for 3D segmentation, in *International Conference on 3D Vision (3DV)* (2019)
27. A. Dai, Scannet: Richly-annotated 3d reconstructions of indoor scenes, in *Proceedings of CVPR 2017* (2017)
28. J. Li, Building and optimization of 3D semantic map based on Lidar and camera fusion. Neurocomputing
29. Y. Li, Deep learning for remote sensing image classification: a survey. *Wiley Interdisciplinary Reviews. Data Mining and Knowledge Discovery*, vol. 8, p. 1264 (2018)
30. C.R. Qi, Pointnet: deep learning on point sets for 3d classification and segmentation. CoRR abs/1612.00593 (2016)
31. C.R. Qi, PointNet++: deep hierarchical feature learning on point sets in a metric space. arXiv:1706.02413v1 [cs.CV] (2017)
32. L.P. Tchapmi, Segcloud: semantic segmentation of 3d point clouds, in *International Conference on 3D Vision (3DV)* (2017), pp. 537–547
33. B. Vijay, SegNet: a deep convolutional encoder- decoder architecture for image segmentation. IEEE Trans. Pattern Anal. Mach. Intell. **39**, 2481–2495 (2017)
34. L. Landrieu, M. Simonovsky, Large-scale point cloud semantic segmentation with superpoint graphs, in *The IEEE Conference on Computer Vision and Pattern Recognition (CVPR)* (2018), pp. 4558–4567
35. C.R. Qi, Deep Hough voting for 3D object detection in point clouds. arXiv:1904.09664v2 [cs.CV] (2019)
36. B. Yang, Computing multiple aggregation levels and contextual features for road facilities recognition using mobile laser scanning data. ISPRS J. Photogramm. Remote Sens. **126**, 180–194 (2017)
37. Q. Hu, RandLA-Net: efficient semantic segmentation of large-scale point clouds. arXiv:1911.11236v3 [cs.CV] (2020)
38. Z. Yang, Std: sparse-to-dense 3d object detector for point cloud, in *The IEEE International Conference on Computer Vision (ICCV)* (2019)
39. Y. Cui, Deep learning for image and point cloud fusion in autonomous driving: a review. arXiv:2004.05224v2 [cs.CV] (2020)
40. A. Zisserman, Very deep convolutional networks for large-scale image recognition. arXiv print. 14 p (2014)
41. O. Ronneberger, P. Fischer, U-Net: convolutional networks for biomedical biomedical image segmentation, in International

Conference on Medical Image Computing and Computer-Assisted Intervention (2015), pp. 234–241

42. K. He, Deep residual learning for image recognition, in *Proceedings of the IEEE Conference on Computer Vision and Pattern Recognition* (2016), pp. 770–778

43. A. Chaurasia, Linknet: exploiting encoder representations for efficient semantic segmentation. arXiv preprint 1707.03718 (2017)

44. M.H. Wu, ECNet: efficient convolutional networks for side scan sonar image segmentation. Sensors **19**(9), 2019 (2009). https://doi.org/10.3390/s19092009

Artificial and Geospatial Intelligence Driven Digital Twins' Architecture Development Against the Worldwide Twin Crisis Caused by COVID-19

Mezzour Ghita, Benhadou Siham, Medromi Hicham, and Griguer Hafid

Abstract

Digital Twins DT have been considered recently as the leap forging technology for digital and physical world fusion. Our analysis to the triple crisis that the world is experiencing as a result of COVID-19 has enabled us to identify several challenges to overcome for current decision-making systems and industrial environments that are constrained to develop adaptative management systems, flexible networks and smart business continuity plans. The impacts of both geospatial and business intelligence as well as advanced simulation have motivated our proposition of a new generic framework based on digital twins to deal with these challenges. Our proposed framework by combining first digital twins with business intelligence tools aims to develop smart information models, value-driven DT architectures and context aware services that tries to mimic real-environment dynamics, and according to its developed smart engines prevent the occurrence of unpredictable events. Secondly, by integrating to this combination, location intelligence provides multi-perspective modelling and geo-statistic data integration with real-time operational data into digital twins offers a generic system for prognostication, optimization and on-line and offline learning. The framework is integrated across the paper within the efforts to deal with the triple crisis of the current pandemic with a focus on Middle East and North Africa MENA regions and particularly Morocco and concretized through an application use case within the field of protective facial masks production.

Keywords

Digital twins • Location intelligence • Artificial intelligence • Supply chain resilience • COVID-19

1 Introduction

Since last December, the world is living under the shadow of a new pandemic, Corona Virus, COVID-19 or severe acute respiratory syndrome 2. This pathogen belonging to coronavirus family has so far induced changes in the socio-economic, geopolitical, technological and demographic profiles of the world's continents, especially the most precarious ones [1]. The undergoing twin health and economic crisis alongside the emerging global dynamics and changing standards in several areas has prompted governmental institutions, and scientific communities to reframe their perception to emergencies responses and policies to meet the challenges of the pandemic and face its health, economic and social fallout and challenges, and supply chains and industrials around the world to reinvent their practices in order to respond to resiliency requirements that the new global dynamic has put in place [2].

Within numerous countries, advanced Information Communication Technologies ICT technologies and digital solutions have been recognized as efficient weapons against COVID-19 [3]. Among these solutions, artificial intelligence and dynamic simulations have been deployed by several countries in order to not only prevent but also counter the global impacts of the pandemic. Artificial Intelligence AI has been used by several countries for early detection of the virus, vaccine trials development, early diagnostic and smart testing particularly within European and Asian countries [4].

M. Ghita (✉) · B. Siham · M. Hicham
National and High School of Electricity and Mechanic (ENSEM) HASSAN II University, Casablanca, Morocco

M. Ghita · B. Siham · M. Hicham
Research Foundation for Development and Innovation in Science and Engineering, Casablanca, Morocco

M. Ghita · G. Hafid
Innovation Lab for Operations (ILO), Mohammed VI Polytechnic University (UM6P), Morocco, Ben Guerir, Morocco

© The Author(s), under exclusive license to Springer Nature Switzerland AG 2022
F. Barramou et al. (eds.), *Geospatial Intelligence*, Advances in Science, Technology & Innovation,
https://doi.org/10.1007/978-3-030-80458-9_7

In the recent years, artificial intelligence has been widely used for industrial processes optimization and industrial value chains resiliency enhancement. Some studies in the context of the current economic and sanitary crisis have explored the potential of these developed results and experience feedback of supply chains smart optimization and its fusion with epidemiological modelling to counter COVID-19 economic effects [5].

Mathematical modelling within the epidemiological field dates back to the 1980s with the emergence of Human Immunodeficiency Viruses HIV. Since its advent, epidemiological modelling has undergone several evolutions and has been able to capitalize on the experience acquired through the various pandemics that have occurred worldwide [6]. Two models have received the attention of scientists around the world during this new pandemic, the Susceptible-Infected-Recovered SIR model, and specifically, the Susceptible-Exposed-Infected-Recovered SEIR model. The first compartment of the SEIR model comprises susceptible individuals who may contact the virus, the second compartment comprises individuals who have contacted the virus but whose viral charge is not yet sufficiently developed to make them infectious, the third compartment represents infected infectious individuals and the last compartment represents withdrawn individuals. The evolution that populations are currently undergoing due to advanced urbanization and various technological evolutions has led scientists to integrate further factors into their perceptions of the different compartments and the existing interactions among them, as instance, urban mobility, locations, environmental and demographic criteria and countries' social and economic indicators [7]. The exploration of these parameters with relationship to pandemic evolution through simulation and machine learning has proved to be an efficient means for pandemic peaks prevention despite being faced with several research challenges mainly for the purpose of this paper.

Q1: *How can we develop a dynamic simulation of pandemic evolution that takes into consideration at real-time spatiotemporal characteristics of the physical environments and population dynamics and the interactions between different economic, social and medical ecosystems?*

Q2: *How can we securely and ethically leverage digital solutions and artificial intelligence potential to build an interactive bridge between different heterogeneous institutions and develop smart and optimal policies that deals with the sanitary effects of COVID-19 on populations and acquire experience feedback for post-pandemic economic ecosystems resiliency?*

As we have been able to see through our analysis of the evolution of the pandemic in several countries and the response and prevention plans that these countries have put in place to face the COVID-19, some of them have stood out by the significant involvement of a collective intelligence not only at the level of governmental institutions but also among local social and scientific actors, hence the important contribution of social, economic and demographic changes in the evolution of the pandemic around the world and for environment dynamic modelling under pandemic context [8]. The recent developments that the world is experiencing with the social distancing measures and prevention policies that countries have been forced to put in place require flexible, resilient and intelligent decision-making and value chain management systems that could adapt their evolution with the progression of the pandemic within a constantly changing internal and external environment, particularly for countries with poorly diversified economies such as the Moroccan country. Collective intelligence in this context can be apprehended through geospatial intelligence that integrates efficiently location intelligence, geospatial data and epidemiological indicators into a unified and distributed platform. Recently, several solutions integrating location intelligence were developed for workforce remote monitoring, supply chains flow management and smart communication between cities' stakeholders and decision-makers [9]. The fusion of these solutions with artificial intelligence-based simulations can be concretized through digital twins. DT are currently leap frogging technological revolution for cyber physical systems within the context of smart factories and smart cities [10]. DT are virtual mirrors of living and non-living entities that imitates these entities' behaviours within their environments by means of digital technologies with the purpose of simulating, modelling and optimizing their functioning and respective interactions.

In the context of this paper as a continuity to our works developed throughout, we explore the potential of a distributed and smart digital twin solution based on location and artificial intelligence for addressing the two research questions identified earlier.

Answer to Q1: *Multi-perspective analysis of pandemic evolution within MENA region under different control and prevention policies with a focus on Moroccan country and its geospatial particularities. Review of the different scientific efforts made up to date to mitigate pandemic impacts on both short and long terms focus on advanced simulation and artificial intelligence.*

Answer to Q2: *Digital twin's fusion with location intelligence for the optimization of supply chain response to different contingencies caused by unpredictable events resulting from external uncontrolled contextual factors and risks as instance the current epidemiological context.*

The remainder of this paper is organized as follows, the second section explores the reversible impacts of geospatial, economic and social factors on COVID-19 within the MENA region with Moroccan country on the loop. The third section introduces a review of the different solution developed by the scientific communities to model and simulate COVID-19, and to mitigate and prevent its economic and social impacts. The fourth section introduces Digital twins' and geospatial business intelligence and their potential for the development of resilient value chains. Section 5 describes the proposal of the DT-Geo-BI platform to mitigate the effects of the pandemic on the value chains of manufacturing plants and to improve their resiliency under multiple unpredictable external risks. Section 6 discusses different application use cases of the proposed solution on case studies within Moroccan context. The last section synthesizizes simulation results, concludes the paper and opens up on future research axes.

2 The World Against a Triple Crisis

2.1 MENA Zone Countries on the Loop

Presently, the world is experiencing one of the most critical health crises in its history, prompting its scientific and industrial communities to renew their crisis management strategies and plans. To date, COVID-19 has caused more than 1,231,615 deaths worldwide, and more than 12,543,694 active cases [11]. Many European countries including France, England and Germany are now facing a second lockdown following a second wave of the pandemic which has compelled them to take more drastic measures to mitigate the spread of the virus and save their health systems from a second surge of overcapacity. The containment measures that pushed countries to close their borders and limit internal and external flows in and out as experienced in the first wave of the pandemic have contributed not only to the disruption of a significant number of value chains for its countries but also for their international partners. This triple economic, social and health crisis is changing the global economic and social dynamics of many countries, particularly developing countries, which are forced to face several challenges that are putting their populations and their development efforts in jeopardy due to economic weak resiliency. In this section, we have decided to focus in particular on some of MENA zone countries. Within the MENA zone, Egypt has the largest population with a population of 102,334,403 inhabitants, Algeria comes in second place with 43,851,043 inhabitants, followed by Morocco in third place with 36,910,558 inhabitants, Tunisia counts 11,818,618 inhabitants, Libya 6,871,287 and finally Mauritania 4,649,660 inhabitants [12].

On February 14, 2020, Egypt was one of the first countries in the MENA region to record the first COVID-19 19 case. Currently, the country has more than 43,398 active cases and 2620 reported deaths. On 02 and 03 March 2020, respectively, Tunisia and Morocco recorded their first cases followed by Algeria on 03 March 2020, Libya and Mauritania. Morocco accounted for a total of 218 deaths due to the virus, Algeria 885, Libya 18, Mauritania 120, Tunisia 50 and Egypt 2620. Figure 1a–c shows, respectively, the evolution of the number of new active cases and the number of new deaths for each of these countries to date and the daily incidence of new cases for the different countries. Pandemic evolution within its countries has followed different paths, which have been strongly influenced by the prevention and crisis management plans put in place by each country and a number of socio-cultural and territorial factors. Pandemic management, particularly in Algeria, Tunisia and Morocco, is characterized by smart use of technologies and scientific research and increased innovation for the prevention of the virus and the response to its health and socio-economic impacts. Tunisia and Morocco were among the countries within MENA region that had best managed the development of the pandemic after a period of containment that lasted nearly a month within Tunisia, and a stringency index for government response tracking that on average had been established at 73.84, and a relatively stable evolution curve of active cases that the country has been able to inflect after a few weeks of strict measures deployment. However, at present, the two countries are registering unpredictable spike in active cases and new deaths that are challenging the sustainability of their pandemic management strategies as well as their health systems' capacities. Egypt, Mauritania and Algeria are experiencing a significant increase in the number of deaths due to the pandemic. The spread of COVID-19 in Libya has added to the political instability and armed conflicts that the country has experienced in recent years, which has weakened the country's healthcare infrastructure. A number of international organizations, including United Nations UN, have been warning of a new outbreak in the country due to the implications of the country's unstable political situation. Algeria is one of the African countries that is undergoing a period of political tension with months of protests and conflicts in the political situation in the country, and with the truce imposed by COVID-19 has been relatively stabilized, however, its social impact is still present and this situation is influencing the development of the pandemic within the country. Studies conducted on the development of the pandemic in Africa had raised the issue of political instability as a factor that could be responsible for further acceleration of the pandemic.

The evolution of the pandemic in Morocco, Egypt and Algeria has experienced numerous peaks and fluctuations which explain the variability in the distribution of new cases deduced from the different peaks present which are due to the appearance of new foci of virus spread motivated by contextual constraints. Tunisia, for its part, has managed to stabilize its curve at some extent during the first 3 months, thus the majority of the points vary around the median except for some points that represent pandemic spikes. However, the country's curve at some extent in the previous months has started to increase rapidly and suddenly, as it is the case for its new deaths curve.

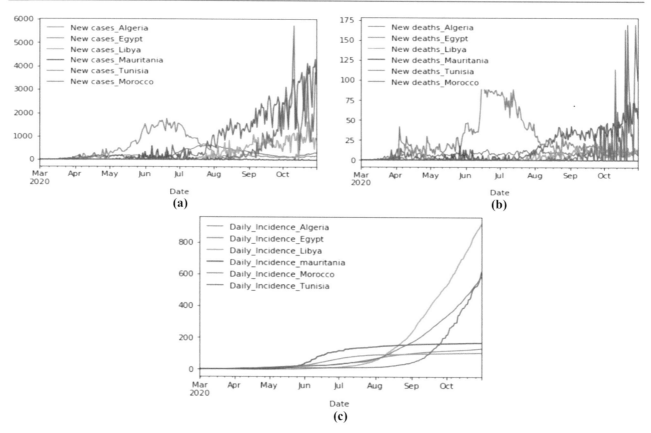

Fig. 1 **a** New cases and cumulative infected cases evolution curve within the different countries from 01 March 2020 to 31 October 2020. **b** New deaths and cumulative deaths evolution curve within the

different countries from 01 March 2020 to 31 October 2020. **c** COVID-19 daily incidence within countries

2.2 Contextual Analysis of Pandemic Evolution: Focus on Spatial and Temporal Variations

Scientific research advances and further development of new tools for data analysis, prediction and applied artificial intelligence have enabled medical fields to strengthen their responsiveness and proactivity to adverse crises. Globalization and increased connectivity of manufacturing supply chains around the world and worldwide territorial changes due to this perpetual evolution have led to a significant urban development. This development was one of the factors that made this health crisis unique [13]. To respond to this new crisis, scientific communities around the world, and thanks to a striking collective intelligence, have succeeded in establishing a new real-time, adaptive and smart system for the monitoring of COVID-19 epidemiological indicators. For the purpose of analysing these policies from different perspectives and as an output for the global community, four main indicators were developed by the scientific group that are economic index, stringency index, government response index and containment health index. Economic index tracks strategies and plans launched by countries in order to encounter pandemic economic impact on all society classes,

such as income support and debt contract relief during pandemic period for persons facing financial difficulties. Government response index is a set of defined indicators that follows different established policies, prevention and response measures in the face of virus spread as well as their resulting impacts on individuals and communities. Stringency index tracks adapted containment strategies put in place in order to stop virus propagation within the country. Those measures include, for example, citizens' mobility management, locally and internationally, and gathering restrictions. Those indexes are each estimated based on defined individual indicators that correspond to different impacts, measures and policies. Ordinal scales are defined for each individual indicator. Government response index is based on 13 different indicators, Containment and health index on 11 indicators, Stringency index is based on 10 indicators and 4 indicators were used for economic index but only 2 are exploited in the context of this paper. In addition to individual indicator scales, a new variable is included for the calculation of the overall value of the various indexes which is indicator flag. Indicator flag determines the geographic scope of the indicators. This indicator shows two states, targeted and general. Health system index include indicators about health system reinforcement and adaptation measures and policies, including awareness and communication campaign, tracing and tracking established systems and testing policies. The development curves of its indicators have been very different from one country to another as a result of a number of contextual factors of a political, economic and socio-cultural origin. Figure 2 shows the correlation results between the indicators that represents the different measures that countries' governments have taken to mitigate the pandemic and the evolution of its incidence, Fig. 3 puts the focus on new cases evolution according to government index and Fig. 4, new deaths evolution according to health containment index that includes testing policies as a main element. Pandemic incidence was calculated based on official open-source data about pandemic evolution within countries as a representative indicator of COVID-19 evolution within these countries. The development curves of its indicators have been very different from one country to another as a result of a number of contextual factors of a political, economic and socio-cultural origin. Stringency index has had a high median for the majority of countries over elapsed period, with a maximum value of 94% recorded in Morocco and a minimum of 84% in Egypt from the beginning of the pandemic to the end of the national lockdown period. Morocco was among the first countries in the region to implement strict preventive measures.

Its response to the pandemic has been characterized by the proactivity of its institutional bodies, which explains its data points recorded over the lockdown 3 months being concentrated around the median which is 66%. This index has known negative correlation with COVID-19 incidence for the majority of the countries studied except for Tunisia and Algeria that represents a weak positive correlation. Restrictive measures have contributed significantly to counter the propagation of the virus. Taking into consideration the cultural and religious patterns of the different countries, this indicator has been, during the last few months, of great importance for containing the virus, especially for countries such as Morocco that have been registering since the beginning of the pandemic numerous familial foci, including as instance during funeral. The Department of Health and Social Care for this purpose in the United States has published a dedicated guideline for funeral attendance under COVID-19 19 circumstances. Through economic index evaluation, we can notice that index correlation coefficients with new cases increase has varied significantly within the same country and from one country to another. Concerning Morocco, the majority of selected points, since the beginning of the pandemic and during the lockdown, are concentrated around the median which is established at almost 85% for lock down period, whereas a few points fluctuate around the minimum corresponding to the beginning of the pandemic and to the period that

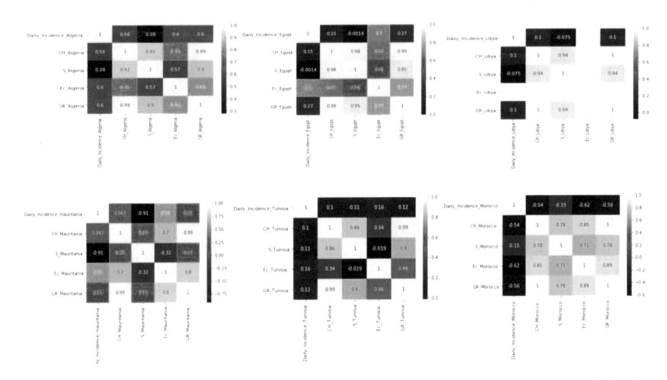

Fig. 2 Correlation analysis of COVID-19 incidence within countries with relationship to countries economic, social, governmental and sanitary response plans

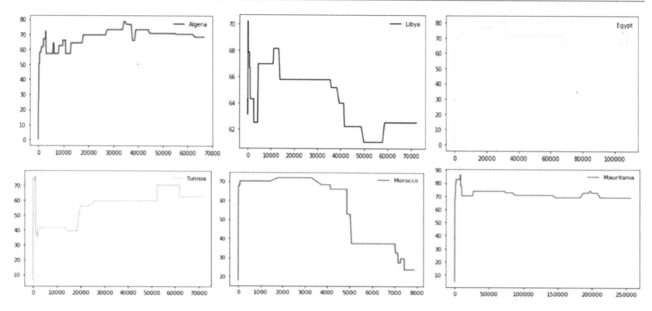

Fig. 3 New confirmed cases evolution with reference to government health index

preceded the containment. During this period, the country has maintained relatively stable economic activity while implementing fairly strict preventive measures, such as cancelling some public events and meetings. Recently, Morocco has been experiencing its second wave of COVID-19, and numerous economic challenges spatially for informal sector and SME's that find difficulties to comply with the restrictive measures imposed by the government which have caused them to be the source of the proliferation of industrial foci. Morocco, as it has been preparing for an economic rebound in recent months, concentrated its economic response strategy on contract and debt relief in particular for start-ups and SMEs as highlighted by many official national sources and through the conclusion of a number of agreements to support local entrepreneurship. Taking into consideration its contextual economic conditions, with a Gross Domestic Product GDP per capita of 3204.1 US$, a relatively high poverty index and a large part of the economic ecosystem in the informal sector, Morocco has established a proactive response plan to the economic crisis guided by a special committee including country's economic agents. Tunisia and Morocco exhibit similar medians, which are explained by the similarities in their prevention and response plans to the economic crisis. Tunisia and Morocco report substantial poverty rates, yet Tunisia has a higher GDP per capita of 3317.5 US$. Libya, meanwhile, is among some of the countries in the world that have not adopted exceptional economic measures in response to the spread of COVID-19. The country's economic index has remained neutral throughout the past period. Of the selected countries, Algeria and Libya have the highest GDPs of 3948.4 and 7686.8, respectively, both

countries are oil exporters. As a result of the health crisis, world oil consumption has fallen by 10% compared to 2019. This decrease is due to the containment and prevention measures imposed to reduce the spread of the virus. The imbalance between supply and demand and the relative fall in prices for countries such as Algeria, Libya and Egypt could have a negative impact on the economies of these countries and an indirect negative impact on importing countries in sectors such as tourism, and in relation to subventions provided to these countries. Egyptian country has adapted a targeted response policy which has involved sectoral financial support. Health system management for all countries selected according to Containment and Health System Index represented variability. Morocco and Tunisia exhibit similar patterns as far as policy implementation and planning is concerned. However, Morocco, contrary to the other studied countries, exhibits a positive correlation of 0.54 between the index and COVID-19 incidence. Egypt is the country with the greatest variability distribution of this index with relationship not only to new cases evolution but also to new deaths increase. Testing strategies within the country and for Algeria as well are not made clear through the data provided unlike other countries, such as Morocco and Tunisia for which we can accurately assess the strategies based on available data from the various local official sources. Testing, as part of epidemiological framework, has a direct impact on the number of cases of contamination detected; it is the main indicator that helps to decrease the number of new infections when correlated with strict quarantine and medical assistance policies. To identify the relationship between testing policies and new deaths, in further parts we focused our study on cross correlation between

Fig. 4 New deaths evolution with reference to containment health index

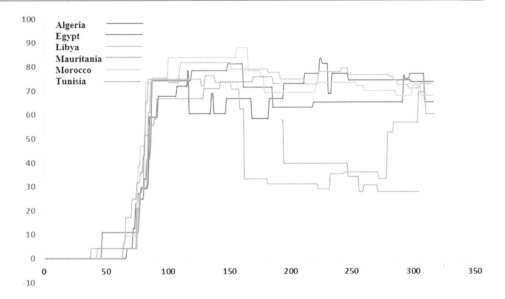

positivity rate and new deaths for Morocco that provides open-source testing data. Government response according to correlation analysis has varied significantly within the studied countries depending on the daily incidence evolution. Faced with the risks for a peak of cases and the different respective correlation with other indexes on their influences on the pandemic incidence, the studied countries have adopted different response plans. Response plans were closely related to the risk of facing a second unexpected wave that results for the country in putting more stringent policies.

2.3 Spatial and Temporal Analysis of Pandemic Evolution and Its Impacts Within Moroccan Territory

2.3.1 Spatiotemporal and Economic Analysis of the Pandemic

The evolution curve of the pandemic within Morocco has undergone a series of fluctuations giving rise to a set of strategies which have varied from a national level to a local level where each region has been treated according to its own curve and its own characteristics and their evolution over time. Figure 5 shows the evolution of the pandemic across the different regions of the kingdom for three periods; the detection of the virus

and lockdown period which runs from 02 March 2020 to 10 June 2020; the period after containment which divided the kingdom into two regions; region 1, where the number of cases had been stable and where the measures were relaxed, and region 2, where the majority of the strict measures were maintained. The second period was characterized by massive testing within industrial environment and the reinforcement of audits for the respect of prevention barriers, including face

covering and social distancing as well as increasing the supplies of diagnostic tests, protective equipment, ventilators and essential medicines for quarantined infected and hospitalized patients. This division with the spread of the pandemic in the course of the months has undergone many changes, and currently, the government has changed its strategy and has put the focus on health systems response by accelerating R&D for vaccine trials and production and treatment development, and the optimization of hospital beds and spaces for diagnostics and treatments and the protection of front health workers [14]. We define the third period that experienced the second wave of the pandemic in the country from 01 September to 15 October 2020. The analysis conducted in the previous sections has highlighted a positive correlation of daily incidence within Morocco with Containment and health index and government response that includes approximatively all the measure that we have cited previously and their impacts. Through this part, we explore this correlation in order to detect the main spatiotemporal features of new deaths and new cases time series within Morocco. The main purpose of this analysis is to define the parameters impacting pandemic evolution with regards to spatiotemporal perspective with focus on critical supply chains within the country. Casablanca-Settat region has seen the largest number of cases in the kingdom; being the economic capital of the country, the region has, according to the 2020 forecasts of the Moroccan High Commission for the HCP Plan, more than 7,408,213 inhabitants with 48% in Casablanca and a density of 18,222/km^2. Rabat Kenitra Salé region with at its heart the capital Rabat of the kingdom counts 4,867,744 inhabitants followed by Marrakech Safi with 4,774,413 inhabitants, Fez Meknes with 4,405,862 inhabitants and Tangier Tétouane El Hoceima with 3,813,854 inhabitants. Morocco has been undergoing in

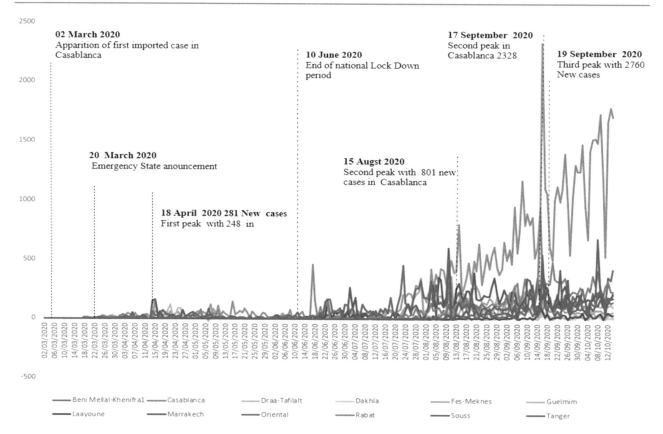

Fig. 5 New cases evolution within regions with a chronology of major events

recent years a genuine demographic transition. Its population is growing, admittedly, but at a slower rate, due to the extension over time of life expectancy in the country, which in 2018 recorded an average of 74 for men and 80 for women, and the decrease in the number of infant's births that was in 2018 17.5 births/1,000 population. Infant, child and maternal mortality rates have also been reduced, thanks to improved health care, nutrition, hygiene and vaccination coverage. In 2018, the rate reached 4.9 deaths/1,000 population, albeit disparities between urban and rural areas and between rich and poor households have persisted. The decline of Morocco's birth rate reflects the decline in the total fertility rate from 5 in the mid-1980s to 2.09 in 2018, due to improvements in women's education levels, increased use of contraceptives and delayed marriage. Young adults aged 15–29 years account for almost 26% of the total population and those aged 25–54 years account for 42.41%, which constitutes a potential economic asset if they can have paid employment. However, at present, many young people are unemployed due to the pandemic, and its impact on the informal sector, which contributes 20% of GDP and offers many employment opportunities for young people, particularly with illiteracy remaining fairly high. Most employed young people work in the informal sector with little security or benefits. From descriptive analysis of regions that

registered new cases, Casablanca is the epicentre of the outbreak within Morocco followed by Marrakech Safi and Rabat-Sale-Kenitra. The pandemic's evolution curve in Morocco and within each region has experienced numerous fluctuations extrapolating the spatiotemporal impact on the spread of the virus and on its sphere of impact, which is increasingly widening from the epicentre to the other regions of the kingdom, putting at risk not only the respective health systems of its different regions but also the resilience of their economic, industrial and social ecosystems, which have been strongly affected by the crisis. Through this paper, we apprehend the co-evolution of COVID-19 with economic, social and demographic characteristics of the different regions spatially, the epicentre of the virus within Morocco Casablanca-Settat region. The progression of the pandemic over time within the population had given rise to a co-evolution of the pandemic with different value chains. Through Fig. 6, we have tried to identify the different medical value chains involved in the pandemic prevention and response plan according to the evolution of the life cycle of the pandemic and the different concerns of the stakeholders for each cycle. We define for each compartment, five main focuses. As instance, the passage from the infected compartment to the hospitalized compartment and the evolution of hospitalized cases with a high frequency as in the

Moroccan case, require the rehabilitation and conditioning of the private and public hospital infrastructures in parallel. This conditioning involves improving the productivity of a number of value chains, such as the production chain for protective equipment for the protection of the medical profession, which has an increasing number of infections and a mortality rate that is undergoing major changes. The geo-statistics of the different regions allow us to identify the resources available and to estimate the shortages in a preventive manner. This detailed analysis should be communicated in full in the three pools that we have defined to represent the co-evolution of the pandemic, the demand and the value chains. Our analysis of the outbreak within Morocco and the different professional foci that occurred through the first 3 months has enabled us to detect the main sectors that has known proliferation of infection cases. Figures 7 and 8 represent these analysis results for new cases evolution, sectorial and territorial during the three first-month period and lockdown period. Data was collected according to Moroccan Ministry of Health daily conference point for epidemiological situation and web scrapping of different trusted e-journals.

Country's evolution curves of new cases highlighted some outlier's data points, a large majority of these points were caused by both some unexpected familial outbreaks and especially professional hotspots within commercial and production units. The evolution of the pandemic in the country has seen similar cases that have climbed the evolution curve to its peaks many times after the country had managed to reduce its national case-fatality rate. These professional outbreaks, which continue to appear in Morocco and around the world, are challenging the response of factories' occupational health and safety systems to the new sanitary requirements imposed by COVID-19 19. Through the figures, we can see that the evolution of the virus has been characterized during the last months by many spikes which reinforce our observation presented on the fluctuations of the real-time reproductive number of the country across regions in the previous section. Numerous spikes have been triggered by industrial outbreaks in particular by the periods of April 20 when the number of cases reached 191 with 95% of cases recorded in industrial outbreaks, May 5 where the number of identified cases was 166 cases with 90% due to professional outbreaks, June 19 where the number of cases increased drastically after a relative flattening out to reach 539 cases caused mainly with 85% by an industrial outbreak and finally June 24 where the country reached a new record of 563 cases resulting in 72% from an industrial outbreak. In order to analyse the effects, impacts and in particular the causes of these fluctuations, we have focused in this paper on two dimensions which refer to the analysis of the different regions and sectors of companies where the cases of contamination have been recorded. Figure 8a shows that Casablanca-Settat region recorded the highest percentage with 44%, Rabat-Sale-Kenitra region followed with 37% and Tangier-Tetouan-El Hoceima region came third with 11%, Laayoune-Sakia El Hamra and Fes-Meknes regions recorded 5 and 3%, respectively. Since 2012, Casablanca-Settat region has maintained its position within Morocco as the region that attracts the most business

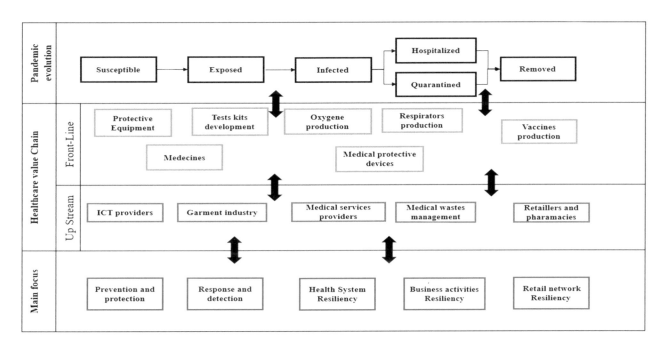

Fig. 6 COVID-19 network mapped to the compartmental model of pandemic evolution

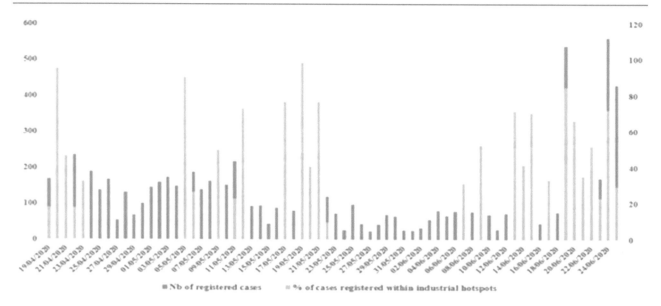

Fig. 7 COVID-19 professional outbreaks evolution within Moroccan country

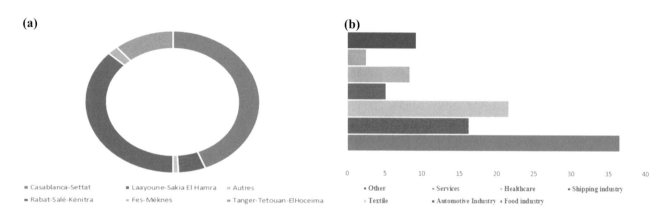

Fig. 8 COVID-19 professional outbreaks distribution according to sectors (**a**) and regions (**b**)

start-ups, the region during the previous year saw the registration of 18,146 new business registrations. In order to extend this analysis to the progression of the pandemic within the region and to detect the factors influencing this evolution, we have carried out a spatiotemporal analysis of COVID-19 within the region. The objective of this analysis is to respond to the first issue raised in the first section Q1.

There are various factors that contribute to the territorial dimension regarding pandemic spread in professional environments, among them the geospatial factor related to both the company and its region, mobility and transport activities within the territory, environmental data and finally economic and industrial activities carried out within those territories. From the presented analysis, we conclude that 37% of the cases were registered at the level of companies in the agri-food industry, 22% at the level of textile companies and 16% of the cases in the automotive sector. The different sectors identified through the analysis present a number of

similarities and discrepancies, each related to a set of factors. Among these factors, the first factor involves organizational practices developed for change management, OHS, ergonomics, risk management and human resources management within industrial sites. The second factor concerns plant infrastructure and related sub-factors, such as plant location, building aeration and workstation layout. The third factor involves the firm as such, its workforce, its strategic business areas and the type of activities in its production units. The fourth factor concerns the company's commitment to OSH and its efforts to improve it and to comply with national and international standards in the field. The last factor concerns pandemic management by the company and its prevention and response strategies. Sectoral analysis of the professional outbreaks highlights the textile and agri-food sectors, both of which are sectors with a strong labour component requiring very strict compliance with barrier measures. The mainstay of the textile industry in Morocco is made up of SME. These

Fig. 9 Correlations analysis between daily maximum and minimum new cases evolution across the country and minimum and maximum distance from pandemic epicentre

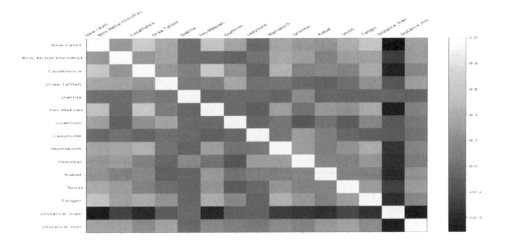

enterprises are at high OHS risks and require a significant investment for the improvement of the office layout. The office layout is one of the factors influencing the spread of the pandemic at the Workplace level. This point has been raised by many organizations, such as the International Labour Organization ILO, with regard to several standards and guidelines dedicated to OSH management at the level of industrial companies [15]. The survey conducted on professional foci once again puts Casablanca-Settat region in the lead. As we have seen from the characteristics of the region, which is the economic epicentre of the country with one of the largest industrial zones in the country in various manufacturing sectors, mainly automotive and electronic assembly, as well as agri-food and offshoring companies. A large part of the activities of the informal sector, which employs a large proportion of the population, is also centred within the region. The economic activity in the region and its various cities and the density of its urban areas have given rise to a number of problems, particularly those relating to pollution and air quality as well as urban mobility. The studies conducted on the correlation of the evolution of the virus with meteorological parameters and climate change aimed at comparing post- and pre-lock down pollution levels have put forward the hypothesis of a potential correlation between climate change variables and epidemiological indicators. In order to apprehend this hypothesis at the level of Moroccan regions and spatially Casablanca-Settat region, which in fact registers a medium air quality according to the AQI index of average leading to the prevalence of different respiratory diseases within the region, we have carried out an analysis modelling the correlation between the evolution of pandemic incidence within the region and the different meteorological parameters, including air quality. Figure 9 represents correlation analysis between new cases evolution across regions with maximum and minimum distance from the epicentre of the pandemic within Morocco. Figures 10 and 11 introduce the results of a Generalized Additive

Model GAM analysis conducted on daily incidence of COVID-19 within Casablanca-Settat, respectively, for both mobility trends and metrological parameters across regions with important incidence. GAM models were chosen for this case study because they can effectively combine different types of fixed, random and smooth terms used in the linear predictor part of the regression model for the integration of different types of impacts. GAMs use smoothing functions to take each predictor variable in the model and separate it into sections that are delimited by nodes, and adjust the polynomial functions to each section separately. Morocco data were calculated for t based on a static baseline of 1 and according to lagged mobility for 2 and 14 days taking into consideration incubation period. From partial dependency curves of mobility evolution with COVID-19 incidence, we concluded that incidence of COVID-19 is positively correlated with workplace mobility, and retail and negatively correlated with residential, grocery and pharma and mobility changes from the baseline. That is, the lower is retail mobility from the defined baseline, due to the pandemic and prevention measures, the lower is incidence.

3 Global Efforts up to Date to Counter the Triple Crisis—Focus on Advanced Modelling, Dynamic Simulation and Artificial Intelligence AI

Since the beginning of the pandemic, the scientific communities have been trying to develop paths to explore virus dimensions, patterns and co-evolution within different countries with different demographic, economic and social parameters and variables as we have seen from Moroccan perspective. The previously developed models for precedent pandemics and epidemics have helped for this purpose, for instance, the different works developed in the case of Moroccan country with SIR [16], SIR with asymptomatic

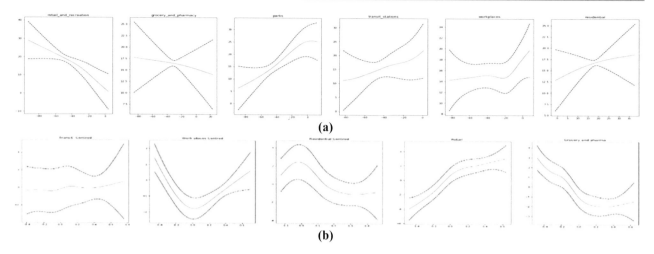

Fig. 10 Mobility lagged evolution with new cases within Morocco (**a**) and Casablanca-Settat region (**b**) for lockdown and post-lockdown period from 15 February 2020 to 15 October 2020

compartments [17] and taking into consideration different control measures [18]. A study conducted in China has shown the impact of urban mobility between cities in the same country on the spread of the virus within the same region, leading to a correlation between the evolution of the R0 of the virus and traffic within regions [19]. This result allowed conclusions to be drawn on the measures to be undertaken to reduce the local spread of COVID-19 after the return to activity and the deconfinement of low-risk areas. Some research communities have been interested in analysing the correlation between prevention and response measures put in place to deal with COVID-19, such as travel restrictions and countries' lockdown measures, and environmental parameters related to climate change and pollution impact on virus spread [20]. Lockdown measures and COVID-19 have enabled to reduce within some countries, including Morocco, GHG emissions, basically NO_2, SO_3 and CO emissions resulting from mobility and some industrial field. The studies carried out on the core of its indicators before and after the containment measures made it possible to draw conclusions on the policies and strategies that the kingdom could put in place to reduce the environmental impact of certain activities and post-COVID-19 good practices to be maintained. The conclusions concluded in Morocco for the actors of the energy sector have pushed governmental and private institutions to strengthen their contribution to the development of renewable energy, off grid in rural areas and the encouragement of research and development in the field. Transportation and aviation are approximately responsible of 13% of Carbon emissions around the globe; lockdown measures, international travel restrictions and social distancing have prompted a lot of countries to reduce their emissions [21]. The characteristics of the affected populations, and the analysis of the correlation between, for example, the age, the gender and previous

health records of the patients have enabled the development of efficient control solutions [22]. Buildings' characteristics, humidity, aeration and workplaces layout are also factors that have been explored. Infrastructure also plays an important role in the respect of sanitary regulations, especially within rural areas that have poor access to water resources and sanitation [23]. Several research communities have worked on the refinement of existing epidemiological models, as instance, SIR model through dynamic simulation of various policies [24, 25], integration of economic networks concerns and changes [26, 27], context Awareness [28] and its fusion with some models, as instance, in [26], poison Markovian process. In addition to SIR model, modified SEIR model was proposed with the integration of new parameters forecasting [29, 30] and real-time estimation methods [31] and advanced simulation and data analytics [32] and machine learning tools and models [33]. Table 1 summarizes these efforts according to three basic approaches that are epidemiological modelling, spatiotemporal modelling, artificial intelligence-based approaches and finally hybrid approaches that integrates both.

3.1 Beyond COVID-19 Towards a New Organizational Model for Value Chain Resiliency

Business resiliency is defined through ISO 22300 by business activity continuity that is described as the set of management practices and implemented policies to identify, analyse and counter the effect of unpredictable incident and threats that can occur during enterprises activities and hinder it from delivering its products and value-added services to its intended stakeholders at their defined constraints and requirements. The crisis we are living is considered within

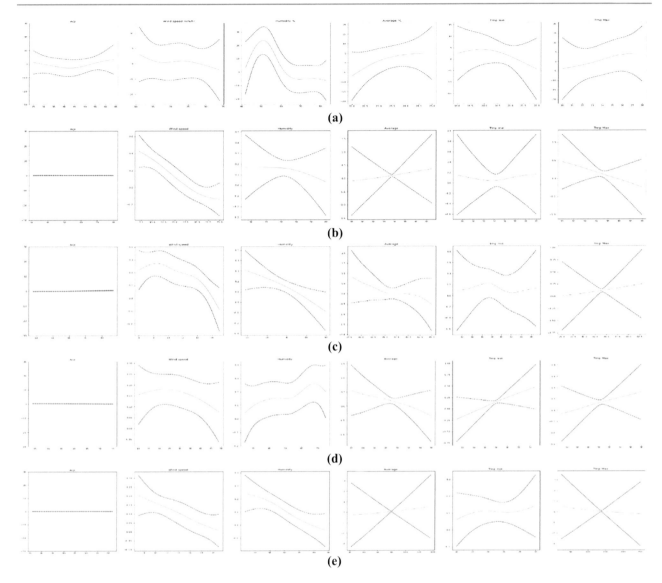

Fig. 11 New cases evolution according to meteorological factors and population levels within Morocco for the epicentre Casablanca-Settat **a** Fes-Meknes **b** Marrakech Safi **c** Tangier-Tetouan-El Hoceima

the standard as one of these incidents. Business resiliency under COVID-19 can be apprehended through different axes we are identifying in the context of this chapter; five of them according to our analysis of COVID-19 developed models and related impacts and risks. The first axis is supply chain management that includes suppliers and suppliers of suppliers' management and consumers of consumers' management with regards to business-critical activities and operations. Supply chain partners cross communication and critical information visualization are defined as a key weapon under the context of the current pandemic. The enterprises should have a clear vision on its network of stakeholders with an emphasis on location intelligence, interactions criticality with regards to disruption threats and network extension opportunities. Regulatory watch,

technological watch and economic watch are three main aspects for supply chain resiliency. The second axis is concerned with IT technology and mainly on three dimensions, cyber security for remote access and for information exchange for both scenarios with home workers and on-site workers, communication performances management from a hardware point of view and a software perspective and finally, data protection. The disruption of public services spatially, communication service due to the traffic and telecommunication infrastructure regionally are also impacts that contributes to this category. The third axis is travel, as travel and mobility are considered as main factors in virus spread, nationally and internationally. Current business and enterprises rely heavily on travels. The risks from travel ban and travel restrictions and the pandemic risks should all be

Table 1 COVID-19 epidemiological modelling, prediction and impact analysis

References	Model	Key features	Simulation	Estimation	Main purpose
[34]	Modified SEIR model	Open-source COVID-19 simulator with a combination of SIR and SEIRD models	Dedicated open-source simulator	Neuro fuzzy inference model for learning and tuning model parameters	The integration of learning through a feedback loop and a fuzzy rule engine
[35]	Modified SEIR model	Development of a SEIRD model simulation that takes into consideration communities dynamics with asymptomatics	Simulation with a SEIRD model with states and conversion method	Individual mobility was integrated by an individual move model that takes four states and control measures were represented through probabilities	Integration of a mobility model for the representation of communities spread probabilities and asymptomatic cases testing of different control strategies
[36]	Modified SEIR model	Simulation and estimation of virus spreading with SEIRD model and Scale free networks	SEIRD model dynamic simulation with power law assumption for scale free networks	Estimation through probabilities of different parameters and definition of R0 as a threshold and two assumptions of free equilibrium or endemic equilibrium	SEIRD model numerical simulation by free scale networks for the definition of spreading networks within communities under quarantine distributions
[37]	Modified SEIR model	Simulation through baseline SEIR model with time-varying parameter for different countries according to control measures and policies	Simulation by the fusion of SEIR Model structure with time-varying parameters estimation	Estimation of transmission rate and removal rates by deterministic spline Ansatz that considers the two ratios as splines with three nodes	Accommodation of SEIR main parameters to time-varying changes of transmission rate and removal rate according to reported data time series and models approximations
[38]	Spatial–temporal analysis	Simulation and prediction of pandemic spread through investigation of its geospatial and temporal patterns locally and worldwide	Dynamic simulation of by SEIR and lag time of individuals movements through truncated power law	Movement estimation by relative attractiveness of locations RA for defining the path of spread according to the probabilities of compartments	Modelling of COVID-19 geospatial and temporal spread BY human mobility simulation and prediction with conformance through government response plans and measures
[39]	Spatial–temporal analysis	Meta-data analysis of spatial and temporal evolution of COVID-19 within patient with different ages and detected virus severity	The exploration of detected patterns for epidemiologic models refinement	Estimation of spatiotemporal evolution of viral detection and viral RNA load for testing locations mainly blood, URT and LR	Spatiotemporal of COVID-19 viral detection and viral RNA load evolution for different classes of patients with different patterns and severity
[40]	Spatial–temporal analysis	The model features a statistical analysis and model of the pandemic local evolution with meteoritical and non-meteorological data and spatiotemporal considerations	Simulation is done at the baseline of a linear model and a Poisson distribution	Effects are estimated based on age demographic characteristics of locations and spatiotemporal meteorological and non-meteorological parameters	The study highlights correlation between meteorological and non-meteorological data and a new model for virus propagation forecasting by a gaussian model
[41]	Spatial–temporal analysis	Simulation and prediction of different classes of SAIR with mobility patterns integration for different regions within Spain	Simulation through SAIR model that integrates Asymptomatic cases as a compartment	Estimation based on mobility modelling through the definition of mobility areas within regions based on provided mobile data	Simulation of COVID-19 taking into consideration asymptomatic cases proportions and country's regions modified mobility patterns due to deployed restrictions across the country

(continued)

Table 1 (continued)

References	Model	Key features	Simulation	Estimation	Main purpose
[42]	Spatial–temporal analysis	Modelling and prediction of pandemic spread through regions by considering virus patterns within their neighbouring and their demographic patterns	The paper do not explicitly define simulation model but gives highlights on spatiotemporal patterns modelling	Use of spatiotemporal models to models virus dynamics with countries. Pandemic evolution is represented by a negative binomial distribution	The developed components of the model categorized pandemic spread by local patterns for specific demographic and correlations and between regions by contagion effect of neighbouring countries for spatial correlations
[43]	AI and ML	Forecasting architecture based on ANN and GWO for testing is applied to COVID-19 time series for the prediction of infected cases evolution	The paper discusses artificial neural networks for COVID-19 complex time series analysis	Estimation of pandemic spread is done based on ANN and hyperparameters tuning and testing are optimized by GWO	Artificial neural networks are used to unease COVID-19 complex time series analysis and to give accurate prediction for peaks detection and preventive response policies
[44]	Hybrid approaches MAS	Simulation of virus transmission among networks of agents defined according to compartmental model	Dynamic numerical simulation of agents states and mobility through the network	Individual agents social force model for the estimation of parameters according to Non pharmaceutical interventions	Evaluation of control interventions on a various scales based on agent networks and different representations of interactions within communities
[45]	Hybrid approaches DT integration	Prediction and simulation of COVID-19 spread across cities through smart cities digital twins for smart environment management	Simulation of cities network and human mobility through cities by analysing feedback of smart city twin	Estimation of climate parameters through artificial intelligence and networks analysis	Smart cities twins simulate cities networks and contribute to decision-making and attributes real-time insights about COVID-19 on different networks
[46]	Hybrid approaches	Prediction and simulation by the integration of a mobility module for pandemic dynamics modelling	Simulation through modified SEIR model integrating mobility in–out flows	Estimation of parameters by an analysis of reported data and their fitting for transition probabilities	The purpose of the simulation is predict end period in order to adapt the measures and to compare modified SEIR model
[47]	Hybrid approaches DT integration	Simulation of different risks resulting from business activities and industrial activities rebound	Simulation of plant and human operators mobility according to various scenarios	Prediction of the risks of social interactions across factories shopfloor and prediction of virus parameters impact on supply chain management	Estimation of risks impacts and the proposition of different adaptation scenarios for preventing and responding to those risks
[48]	AI and ML	Forecasting of virus parameters and variables for USA and India by deep learning algorithms for predictions LSTMs	The implementation of the model provided can serve dynamic simulation	Forecasting of pandemic spread is based on the predictions of deaths and infections by a comparison of RNN with three LSTMs	Providing accurate predictions of the virus evolution within countries with different demographic, social and economic characteristics
[49]	AI and ML	Testing of different regression models for pandemic forecasting within Egypt	The paper discusses key regression models for mathematical modelling and simulation	Estimation and forecasting of is based on the predictions made on the basis of regression models and time series analysis	Regression models for virus predictions can help predict peaks and pandemic end and contributes to the establishment of control policies
[50, 51]	AI and ML	A multilayer perceptron architecture is proposed and applied on time series data of COVID-19 for forecasting and spatiotemporal patterns detection	The paper proposed a learning and prediction architecture based on MLP without explicit simulations	MLP is used for prediction and multigrid search for learning and optimization of architecture hyperparameters	The estimation and forecasts obtained from the model are relevant for policies definitions and evaluation, peaks prediction and identification of virus spread patterns across different regions

taken into account and their impacts managed through crisis plans. One of these plans include the management of incoming flows of travelling employees and recommendations and measures to be communicated to the staff. The fourth axis concerned with Occupational health and safety system resiliency with regards to each pandemic OHS risks inside the workplace and business activities productivity constraints. This axis includes to cite a few security incident management, social distancing, testing policy, professional outbreaks detection and prevention measures, EPI management and requirements with regards to basic ergonomic needs of on-site workers. Standards and recommendations developed in the context of COVID-19 nationally within Morocco and internationally for the development of this axis are reviewed through Table 2. The last axis is interested in the core of the business management continuity plan. This last element is the most important one in having to be prioritized as it's a knowledge experience feedback repository for overall resilience of the business itself. Evaluated based on activities mastery and stakeholder's exhaustive definition of needs and their mapping to main activities and operations of the company. The exploitation of risks management basically through ISO 31000 can structure risks identification, analysis and business continuity plan and strategies implementation when the incident occurs.

4 Location Intelligence and Digital Twins State of the Art of Existing Solutions and Their Potential for Economic, Social and Industrial Resiliency

4.1 Digital Twins' Concept, Applications and Paradigms for Implementation

Digital twins were defined according to ISO Standard digital twins in manufacturing as virtual replications of real systems and assets in the real world capturing its data and functioning within the real environment for optimization, tuning, quality evaluation within manufacturing. For the industrial internet consortium, it is identified as the standard and formal representations of non-living assets that contains its main information and mimic its behaviour patterns within different contexts in order to communicate, store and analyse it according to different viewpoints perspectives [52]. From that deals with digital twin application in healthcare with regards to ISO/IEEE 11073 standard for digital healthcare applications, it is digital replications of living or non-living assets featuring censoring capacities that mimic living and non-living entities perception mechanisms within the real world and employing it in order to feed virtual models contained in the cloud and as inputs to artificial intelligence

Table 2 Standards developed for the development of OHS system

National preventive plans and measures	National barrier measures within facilities
– The establishment of new management and organization guidelines and protocols for industrial and commercial units in times of crisis – The reduction of workforce within manufacturing plants and management of social distancing – Analysis of risk scenarios and implementation of preventive response plans – Creation of monitoring and audit committees for compliance with recommended measures – Implementation of renewed security policy based on internationally agreed principles in response to COVID-19 risks such as 4M and 3C	– Strict regulations and penalties for violations of the measures – Social distancing regulated by control and monitoring mechanisms implemented in professional environments – Provision of the necessary Personal Protective Equipment (masks, hydrological gels, disinfection zones, etc.) and the reinforcement of signaling at industrial sites – Redesign of workplace according to the new standards (workstation separated by physical barriers and Plexiglas)
International preventive plans and measures	International barrier measures within facilities
– Update to the ISO 45000 standard for health and safety in the workplace – Open-source publications of good practice guides and protocols to be followed in the workplace – Simulation of risk scenarios for social distancing – Training of operators through virtual and augmented reality-based workshops – Minimum physical contact through the reinforcement of production systems compliance with Lean Manufacturing principles (5S, Kanban, 5 Zeros...)	– Work environment redesign and implementation of thermal cameras and reinforcement of Internet of Things deployment for health and safety monitoring at manufacturing plants – Teleworking and re-organization of labour – Limitation of physical contact through enhanced task automation, dematerialization, facial and voice recognition techniques, and smart warehousing and inventory management

engines for different purposes and based on numerous tools for data analytics, prediction, optimization and learning [53]. The core of its definition are real and simulated data, communication between real and virtual spaces and information models of the systems. All these definitions and our previous explorations of the concept agrees on three common dimensions for digital twins' development that are models, data and services. Models constitute the virtual abstraction of the real counterpart according to different perspective, domain-oriented representations and value-based structures. Models are defined as instance for digital twins' applications within factories according to the three-dimensional representation of asset within the real world that gives it three space dimensions and one temporal dimension referring to assets' evolution throughout their defined life cycle phases. Space dimensions refer to asset horizontal and vertical integration according to factories equipment hierarchical architecture, whereas the vertical integration is based on the layered functional architecture of the enterprise that connects office floor actors to shop floor actors. This digital modelling abstraction of physical assets gives births to different data and knowledge classes and categories for digital twins that were defined through three blocks by field practitioners. They are master data, operational data and historical data. These three blocks constitute digital twin data architecture and information viewpoints. Master data are linked to digital twin's early development phases, as instance, CAD documents, manufacturer specifications, requirements referential, testing results and all data sources that defines the conceptual structure of the asset. Operational data are usage and maintenance data that are generated through asset integration within the field, and as a result of its interactions with its physical environment and stakeholders, it can be quality data, production data, environmental data, geospatial statics. It helps to evaluate asset efficiency with regards to its environment through enterprise spatial dimensions. This operational data gives birth to historical data introducing thus the temporal dimensions to this spatial data. Models helps to contextualize and understand these three categories of data and to build valuable information models as well as to exchange them with multiple parties that consists of factories networks. The integration of all these sources and their understanding through information models are sources that fed the multiple service and purposes that can be attained by integrating a digital twin into a factory decision-making architecture and management systems. We can categorize digital twin services into three categories according to their founder definition; these three categories are interrogation, prognostication and learning. Interrogation places digital twin as a virtual watch and digital replicas for all the systems and systems of systems within the real world. The second service is prognostication that represents prediction about future states of the real environment based on communicated

interrogation results. Prediction tools that are mainly based on classical and distributed artificial intelligence tools can help to discover hidden features contained in operational and historical data. It works as a proactivity module integrated in the virtual replicas of the plant and taking advantages on both real data and simulated data. The last axis learning is apprehended by different tools that integrates physical modelling strength with data analysis and artificial intelligence methods and algorithms.

4.2 Geospatial and Business Intelligence for Context Aware and Smart DT

Geospatial intelligence was defined as a knowledge and information-based domain consisting of analysing human activities on the earth according to spatiotemporal dimensions developed as a result to human contextual dynamics and spatial characteristics in order to support multiple decision-making processes [54]. During the recent years, the field has been gaining a lot of attention due to the large set of developed technologies such as Geo-information systems GIS and Business Information management that are currently contributing to the development of numerous sectors mainly urban management, particularly human mobility management and construction sector. Population dynamics due to the different technological and industrial revolutions and the increasing pressure that they are both applying on earth's natural ecosystems are compelling today's supply chains to take steps for new organizational structures and approaches that includes a broader management horizon taking as a feature both temporal and spatial dimensions for systems context awareness development. In the context of this chapter, context awareness is defined through a three-dimensional representation of the system, integrating system temporal view and evolution through its lifecycle, systems horizontal evolution with regards that its position inside enterprises equipment's architecture and finally systems vertical integration through enterprises functional view. Digital twins as defined earlier are the digital replicas of the systems in the digital world thus it has to accurately represent their real counterparts respecting this three-dimensional view. Geospatial intelligence contributes to the integration of horizontal and vertical dimensions into digital twin's development process but also with regards to its implementation with the physical environments.

Multi-dimensional representations: Complex systems operational status depends closely upon their geospatial context of evolution. The understanding of this context and its integration in the analysis of system performance changes through interactive mapping allocated by Geographic Information Systems Technologies combining geospatial analysis and business intelligence will be of great added value for new digital twins' generation.

Geo-statistics: There are several dimensions involved for the improvement of industrial complex system performances. Among these dimensions, system spatiotemporal features can be integrated through contemporary tools of geospatial intelligence. Digital twins aim at acquiring and analysing field data and systems-related knowledge in order to provide real-time decision-making support. The fusion of GIS systems analysis and digital twins' simulations results presents a great advantage for complex systems smart monitoring applications.

On-line Monitoring and Secure Networks: Parallel to the development of the digital twins was the development of another concept referred to as digital thread which aims at establishing digital systems engineering that would digitalize product manufacturing lifecycle and thus make complex product engineering flexible, efficient and connected to an extended network of stakeholders and spatial temporal constraints. By merging this lifecycle with advances in geospatial information systems, we can revolutionize complex systems design. The last axis which concerns lifecycle representations within the virtual world was explored by a lot of researchers in the field that highlighted the importance of digital modelling and value-based business model development. Digital modelling and lifecycle phases simulation requires the integration of multiple data sources and information models that are fed with three data categories identified for digital twins that are master data, operational data and historical data. Operational data mimics the multi-perspective view of the system, and as a result, the representation of its different values for different enterprises stakeholders. The integration of all of this models which made up the holistic vision of digital twins requires performant technologies and approaches that merges smart data analysis and visualization.

Business intelligence consists of planning decisions and defining business models and values according to an in-depth and smart data analysis, modelling and visualization based on the integration and processing of heterogeneous value chain data and information in order to extract important hidden insights valuable for stakeholders' requirements fulfilment. Business intelligence needs the acquisition of multiple data from heterogeneous sources during different timespans which requires important data storage, processing and valorization infrastructures that can be enabled by the cloud. Business intelligence helps for the development of smart value chains, smart processes and through the integration of geospatial intelligence smart products that are cautious about all the multiple spatiotemporal dimensions of the complex physical environment.

5 DT-Geo-BI Platform for Smart and Resilient Supply Chains

Our vision for the development of smart value chains, smart processes and through the integration of geospatial intelligence smart products that are cautious about all the multiple spatiotemporal dimensions of the complex physical environment was tailored through our previous work in the scope of the conference whom this chapter is an extended version by the proposition of a DT-Geo-BI platform [55]. The basic building layers of the platform are context layer, cognition layer and users' layer.

Context layer is constituted as we have detailed in our previous paper by different heterogeneous data sources that helps to identify, define and tailor system representations in the virtual world. Our in-depth exploration of BI has motivated our proposition for the combination of data in motion potential with data at rest potential to constitute the building blocks of this layer and as a result how its different output will be organized. Data in motion are the set of acquired stream data and information in real time from the field that depends on systems operational data with the integration of geospatial statics that we identify through an integrated GIS module through the platform. The contribution of GIS systems relies on enhancing the contextual identification of assets in the physical world towards its twin representations across the digital world. The aim of acquiring this data in real-time manner depending on process-imposed data processing delays is to enhance the spatial identification of assets and to give significant insights on its development according to its hosting environment. The second block that is constituted by data at rest producers includes as we have defined before in the conference paper all batch processing data that includes master data, historical data and all data sources that are related to real counterparts' conceptual structure and functioning through defined periods of time and domain-based information that is acquired from involved parties in the physical environments of the real counterpart and distributed within different sources of information and through management systems that contributes in the functional layer of the architecture or that can be acquired by scrapping of external sources.

Cognition layer has as a purpose to contextualize and analyse the gathered data by feeding digital models and replicas of the real counterpart at different scales and levels. Two blocks constitute this part; the first block is data acquisition and data processing. Data acquisition is realized through two functions, spatial recognition and data contextualization, and building of contextual environment. Spatial recognition and data contextualization consist on giving sense to the acquired data through combining it with domain knowledge and value defined features and indicators.

Building of contextual environment is done through the combination of this indicators with the models to gain insights on the evolution of the systems within their physical environments and with relationships to stakeholders needs, requirements and constraints. GIS systems contributes as a support for enhancing the connection between different networks of digital twin agents and their produced insights about their real counterparts. Data processing and analysis consist on the extraction of knowledge and the identification of patterns from the results of the simulation on data in motion and at rest for the definition of new business and monitoring rules. Analysis take advantage from advanced artificial intelligence tools for prognostication, optimization and learning and combine the resulting outputs with the real-time simulation of processes in order to give a proactive perspective to the systems that constitutes the real environment. The resulting knowledge is then stored and shared across different knowledge repositories responsible for new rules generation and serving the tuning of off learning mechanism constituting a smart support system for the platform and an automatic recommender for decision support.

Users' layers is revived with the incoming information and knowledge flows resulting from cognition layer and with the different services expressed by users' interests in the platform. Different services constitutes this layer, as instance, Geo-statics at different levels, plant level, system level and component level, Map-driven dashboard that are based on a comparative geospatial analysis of systems evolution within their environments, digital assistance service that gives access to different knowledge acquired about the asset through the cognitive analysis, prognostication and learning results and last but not least, performances analysis through business intelligence as an output of data understanding across various business and domain-based models. The fusion of Geo-BI technologies and digital twins can result in the development of resilient supply chains and decision-making systems. In order to map this new three perspective vision to current context paradigms, we have proceeded to a mapping of the different proposed solutions of digital twins particularly those based on artificial intelligence and involving Geo-BI for dealing with the five axes we have defined as major resilience aspects under the current context, the focus is put on only four of them. Table 3 represents this review, and Fig. 12 details architecture components in the context of a manufacturing company supply chain network.

6 Application Use Cases and Discussion

Our in-depth exploration of virus reversible impacts on countries supply chains and the spatiotemporal analysis of the virus has highlighted the importance of geospatial modelling and simulation fusion of the virus with value chain models and simulations for a proactive response plan but also for resilient supply chains in the post-pandemic future. In Sect. 2, we tried to explore by a parallel mapping of virus propagation compartmental model with supply chains response and human mobility dynamics as one of the decisive factors of pandemic evolution and local and national propagation. The proposed use case that depicts two potential application is extracted from the identified supply chains as a proof of concept of our DT-Geo-BI platform. The second application is identified through our analysis of professional outbreaks within Moroccan country during the first 3 months of the pandemic. We tried to deal with two of the proposed axes through our previous sections that are travel-combined with supply chain management and Occupational health and safety system resiliency.

6.1 Use Case—Occupational Health and Safety System Resiliency Within Facial Protection Masks Value Chain

Description of the physical twin and its environment and the motivations of the use case

One of the most important value chains within the scope of the current pandemic is that of protective masks. Communicating with different stakeholders and subject to several technical constraints and quality and productivity requirements, this value chain brings into play different actors and areas of resilience. Among the end customers networks of this chain are healthcare workers, who are one of COVID 19's most exposed frontline workers. The second customer network is constituted by industrial and manufacturing companies who's required to ensure the protection of their work force. Among the end customers networks of this chain are healthcare workers, who are one of COVID 19's most exposed frontline workers. The second customer network is constituted by industrial and manufacturing companies who's required to ensure the protection of their work force. The last customer network includes individual consumers through supermarkets or retailers and pharmacies. The world has taken a particular interest in this value chain, and in

Table 3 Mapping of Geo-BI-based DT solutions for dealing with resiliency	Resiliency	Geo-BI DT platform potential	AI technologies contribution
	Supply chain management	– Suppliers smart management through location intelligence – Business intelligence for the selection of suppliers' alternatives and its combination with the first location intelligence – Digital twins plant simulation offline for simulation of raw materials and inventory management for order predictive scheduling – Drawing of customers profiles and predictive analysis of market demand and as a result optimization of supply chain response – On-line digital twin simulation of plant logistics and real-time tracking of assets within the plant through location intelligence – Costs management across supply chain through business intelligence and simulation of costs for different scenarios implementation of scenarios within DT	– ML for time series analysis and demand forecasting – Multiagent systems for optimization and smart networks for resources management – Optimization algorithms for network and distribution management – Visual recognition and object detection for quality check and inspections within manufacturing thus supporting manual inspections and human contacts – Game theory for finding equilibrium between different policies as instance supply chain goals management with respect to sanitary goals
	IT infrastructure management	– Security risks management by network twining and security and interruptions risks impacts simulation through digital twins – Location intelligence for the management of workers networks working from home workers and on-site workers and their interactions – Telecommunications disruption risks management and infrastructure evaluation and inspection for home working and onsite remote working through business intelligence and location intelligence by GIS as instance	– ML for risks detection, classification and prediction – RL for cyber security and multiagent systems for networking simulation reinforcement – Expert systems for recommendations and BI results management and smart exploitation – Adversarial Networks for cyber security through DT simulation and AI
	Occupational health and safety system resiliency	– Mobility simulation and tracking across workplaces through location intelligence and digital twins – Office layout management and optimal configuration of workstations and shopfloor through business information models driven digital twins – Ergonomic monitoring with workplaces under pandemic conditions and social distancing measures by smart metering and DT – Propagation risks management within workplaces for areas of major safety risks – Lean practices reinforcement by plant simulation and location intelligence as instance smart tracking and tracing of assets for smart storage and Kanban – Management of co-existence of robots and human workers within workplace	– Cognitive sensing and embedded artificial intelligence for smart communication and edge analytics – ML for risks estimation, classification and prediction – Virtual reality and augmented reality technologies – Expert systems for knowledge capitalisation

(continued)

Table 3 (continued)

Resiliency	Geo-BI DT platform potential	AI technologies contribution
Business management continuity plan	– Threats and opportunities management through updated economic watch and technological watch reinforced by GIS functionalities and predictions results through digital twins – Integrating ISO 31000 for risks management through evaluation parameters management by artificial intelligence estimations and testing of sequences of risks on the plant digital twins through its composite level – Knowledge feedback capitalization within accessible and flexible knowledge repositories	– Expert systems for smart recommendations and crisis plan real-time update and enhancement – Predictions algorithms for risks evaluation including neural networks

particular, Morocco, which is today not only a producer-exporter but also a manufacturer of mask conception machines throughout their categories, materials 3D printing technologies included. In order to better understand the value chain, we have modelled it according to a modified BPMN model. The production value chain involves three major stakeholders. The first network consists of raw material suppliers, for the purposes of this paper, we will only consider suppliers of reels and raw materials involved in the assembly process. The second block is constituted by the manufacturing network. This network includes a number of processes that we divide into two parts, shopfloor and office floor. The last block is the consumer network we defined earlier and whose demand varies considerably as we have seen in the second section of the paper. Manufacturing company pool is constituted through office floor for business operations management and decision-making and shopfloor for operational control. The current control flow of the manufacturing process is triggered by raw material reception according to customer order and stock availability inventories. Problematics that can occurs at this level are related to suppliers' networks availability. Three tiers of suppliers are actors within supplier pool that are material suppliers, service providers that depends on company processes control structure and the last tiers equipment manufacturer that comes to be an important actor within the value chain of companies' resiliency system under the current context.

Raw material welding process: The welding of the masks is carried out on an ultrasonic welding machine which processes the received coils to give shape to the constituent layers to give the final structure of the masks. This first process is our first robotized station. The first operation at the station is the superimposition operation which consists of superimposing the 2 or 3 layers of material chosen for the design of the masks and their central folding. This second operation is followed by an insertion operation which aims at inserting the nasal support, followed by the folding operation of the upper and lower parts, then the welding of the upper and lower folds and finally the last operation is automatic cutting of mask contour.

Inspections and Quality control: Quality inspections are an essential part of the manufacturing process. Quality inspection is done in two stages: a quick inspection for the control of visually detectable defects followed by a longer inspection which consists of performing several quality tests to validate the conformity of masks to various international recommendations and national standards including the standard designed by Imanor in Morocco NM/ST 21.5.201-2020. We cite, for example, among these tests, barriers testing of materials for bacterial filtration rates, physical testing and safety testing for microbial cleanliness, biocompatibility and flammability according to ISO 10993. Quality tests are performed both by human agents which is the case of our detailed use cases and by dedicated machines that takes advantage from visual recognition techniques for material structure quality control. Quality parameters were resumed by professionals of protective face masks evaluation as the 4 F that are Filtration, Fluid resistance, Features and Fit.

Masks assembly process: Assembly is based on four welding operations or two depending on the equipment used. This process for our case study is one of the most labour-intensive processes as the operators are responsible for performing all four operations for each batch of incoming masks. The number of masks for each outgoing batch moving on to the next process is 100 Mask.

Masks quality inspections and conformity control: Each batch of 100 outgoing products goes through a quality control which allows to evaluate the conformity with the references. This process is carried out by experienced quality

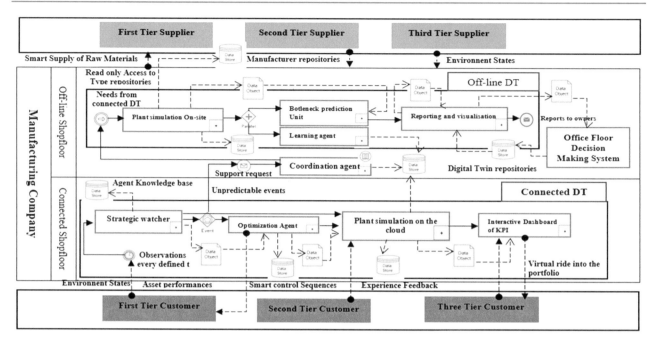

Fig. 12 Geo-BI DT architecture structure example for manufacturing company

control operators. Two outputs result from this process. Compliant masks are routed to the followed process, non-conforming masks are sent for reworking or they are classified as non-repairable and sent to further analysis processes according to their non-conformity indexes. The outgoing flow from this process is then redirected into sterilization station. The output batch for each sterilization cycle is 10.

Bagging, packaging and storage process: The last operation within the production unit is packaging, which is currently done manually by the operators. The outgoing mask packs are then stored in a conditioned storage warehouse within the unit and prepared for transport and distribution according to order schedule communicated by the office floor.

Description of the digital twin and its main modules

The analysis of the real system from a broader perspective allowed us to detect the major interactions of the real agent with his real environment that the twin will have to replicate through his abstraction of real-world entities. We thus defined three categories of actors that are personnel, equipment and material. The proposed information structure is based on recommendations from ISO 23247 part 3 for digital twin in manufacturing that deals with the digital representations of manufacturing elements. In the context of this paper, we put the focus on personnel class and material class. In the context of the current pandemic, occupational

health and safety has become one of the priorities for manufacturing companies. Simulation has been used previously in several research for OHS risks management and digital twins as a tool for advanced simulation; recently, some research works have proposed the integration of digital twins for physical ergonomic and operational risks management in the work place. The contribution of the proposed use case is the integration of business continuity plans perspective within this vision through artificial intelligence and GEO-BI integration within the replication of operators within the digital plant. The structure of these three types of categories are defined through Fig. 13.

Focus on Geo-BI contribution

Through the descriptive diagram of the different stakeholder networks and the production chain, we can see the problematic areas of the chain where it would be relevant to integrate an external virtual support. Two networks are clients of the GEO-BI module of the DT platform. The first client is the Customer Abstraction Loop in the virtual world, the GEO-BI platform monitors suppliers and the development of client's profiles. For each series of product that are facial masks for our company, there is a distribution point that expresses customers' needs but also gives hints through historical data and spatiotemporal analysis about demand evolution across the consumption networks. Our goal is to catch these needs and come up at each external and internal environments state with new adapted production sequences and management schedules. From this perspective, both distribution points and

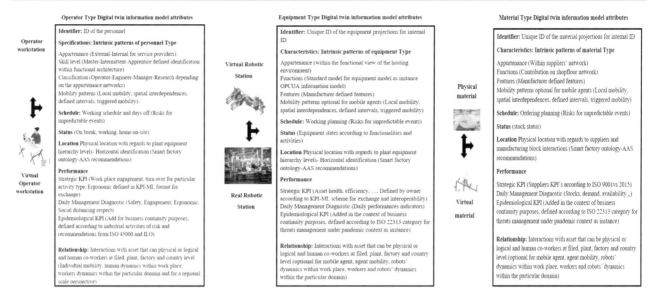

Fig. 13 Digital representation information attributes for personnel type digital twin (**a**), equipment type digital twin (**b**), material type digital twin (**c**)

producers of the company's internal structures are assets that communities through their respective digital twins in the virtual environment by dedicated agents try to learn from each other's experiences and action policies according to a distributed architecture preventing both customers' disruptions and producers wastes mainly over process and over production. It also contribute in enhancing external ecosytem response to the evolution of the pandemix around the world, and to threats of pontential disruptions in raw material supply. In compliance with ISO 9001 version 2015, a list of suppliers is established and registered; according to this registration or according to the outputs of the company's resource management planning, ERP material type Digital Twin is fed. Through the GEO-BI interfaces of each material type, the relevant users of the value chain can assess the demand from the customer network to forecast fluctuations, seasonality and check the availability of its suppliers from a multi-perspective view of both networks.

Digital physical twin connections

The twin main conceptual structure that is represented in the previous section is mainly based on two main module that are connected DT and offline DT.

Connected DT is based on three units that we defined as agents as the aim is to tailor their conceptual structure through testing and simulation of the architecture to include more advanced communication, processing and learning capabilities of autonomous and smart agents.

Strategic watcher: The first agent in this use case is the strategic watcher that communicates with the filed in order to

detect potential threats and to learn from current operational threats to prevent future critical risks. The agent in this particular use case is provided with observations from assets in he real world as instance their states evolution according to changes in the physical environment resulting from both suppliers and customers dynamics and epidemiological parameters evolution. Each physical agent represented by the operator type digital twin information model attributes is assigned a number of threats. Threats are related to agent functions within the shopfloor, the impact of him producing the desired output and its different interactions within the environment. Interactions are defined across the enterprise functional view. The definition of this threat results in the evaluation of a major index for occupational management resiliency that we define as Action Priority Rating APR. Similarly to Risk Priority Index for risk management, in the context of business continuity, APR is defined as the priority index of actions to be taken in response to a critical incident. This index enables the integration of strategic threats management across enterprises functional levels integrating as an added dimension human agents' issues in the overall control loop of shopfloor. The watcher has, as a goal, to observe the environment and prevent the root causes of the probable threats.

Optimization agent: The optimization agent receives the messages communicated by the strategic watcher agent based on the proposed action plan, its feedback from the environment and its knowledge base performs new sequence proposals and conditions under which the actions proposed by the watcher agent are to be deployed. This agent has the ability to communicate with a replica model of the real

(a)   **(b)**

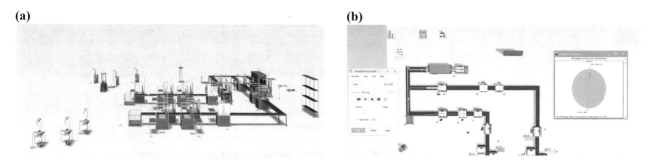

Fig. 14 Virtual factory offline twin layout (**a**) Analysis of workers efficiency and part flow across production line through Sankey diagram (**b**)

system to test these results in real time and improve its decision-making process by testing different resolution approaches through the exploitation of distributed cloud capacities. In this particular use case, the agent helps mitigate thresholds identified by the watcher and proposed actions suggested by the watcher for operational optimization of targeted processes by the incoming predicted threats. The aim from the optimization is to ensure activities continuity under the identified threats and according to real-world constraints at instant t and t-1.

Connected virtual plant: This representation of the system serves as a dynamic environment for the agent to test their scenario and to learn from their failures; it's a major component in the loop as it receives real data to populate its model which can be more significant for critical application that cannot tolerate uncertainness and contingencies resulting from uncomplete or delayed stream data pipelines with the field. This first virtual plant model is constituted on the basis of our analysis of the different processes and their interactions as well as real operators' availability and existing schedules that we aim to tailor in the context of incidents and unpredictable events occurring as a result of environment states changes.

Interactive dashboard: Its represents agents communication interfaces with the environment and its decision makers. Its provides decision makers with an opportunity to firstly contribute to platform processes and secondly to enhance agent's flexibility. Users can test their own scenarios on the environment after getting insights from the observation's agents and the API indices to see the impact of their decision on the plant through the virtual plant.

Coordination agent: This agent is triggered by the occurrence of unpredictable events with critical API that needs actions that cannot be executed on the virtual instance environment and needs more in-depth analysis. The goal from developing this agent is to create a communication interface for synchronization between the twin instances and their types at particular situations. Coordination agent is provided with a learning unit in order to adapt to all kind of situations and to get more into the complexity of both instances and types.

Plant simulation onsite: This offline simulation serves as a virtual shadow of the plant that works continually to perform new scenarios and to detect bottlenecks within the real environment with the provided information from users without interrupting the analysis done at the connected level. We have adapted the simulation environment to the real plant as a first version within Plant Simulation Software. Figure 14 introduces the developed simulation and the first primary results based on current organization and the implementation of real plant characteristics.

Block necks prediction and learning: These are based on the analysis of different scenarios under the context of the different threats and operational risks especially the ones that can occurs as a result to COVID-19 and social distancing constraints. At a large scale this simulation will be adapted to more broader situations that hunters' operator's safety and well-being. Ergonomic quotation being a major tool for this analysis will be develop among others indicators that are classified into categories defined as security, engagement and ergonomic. As it is an offline simulation this provides broader opportunities for testing, validation and learning.

7 Conclusion and Future Research Axes

The analysis we conducted on the spread of COVID-19 throughout the world has enabled us to draw a new perspective for the exploration of pandemics which focuses on the respective co-evolution of the pandemic with the dynamic social, economic, environmental and industrial changes of the populations. This point, although having attracted a large community during the previous pandemics that the world experienced, has not been able to stand out until recently, thanks to the evolution of communication and information means and technologies which increasingly allow us to explore the hidden patterns of many complex natural phenomena. Throughout our paper, we have tried to focus on these triple aspects by the definition of new resilience axis for industrial field. The chapter addresses these proposed axes by main contributions as follows.

– Review of different spatial–temporal, AI based models, and hybrid models for epidemiological modelling and pandemics impacts management, and Spatiotemporal and contextual analysis of COVID-19 19 within Moroccan country.
– The proposition of a generic solution combining advanced simulation and multidimensional modelling through DT, geostatistical exploration through location intelligence and new business models by BI for strengthening the resilience of critical value chains.
– The tailoring of a new resilience perspective that integrates artificial intelligence tools for adaptative value chains driven by flexible and cooperative decision-making system. The vision developed throughout this chapter could also support the implementation of the new global dynamic that puts at the heart of its interest the preservation of humans.

Acknowledgements Work carried out within the framework of the Cooperation Agreement for technological and scientific development concluded between the UM6P and the FRDISI.

References

1. M. Buheji, K. da Costa Cunha, G. Beka, B. Mavrić, Y. Leandro do Carmo de Souza, S. Souza da Costa Silva, M. Hanafi, T. Chetia Yein, The extent of COVID-19 pandemic socio-economic impact on global poverty. A global integrative multidisciplinary review. Am. J. Econ. **10**, 213–224 (2020).https://doi.org/10.5923/j.economics.20201004.02

2. C.M. Jacob, D.D. Briana, G.C. Di Renzo, N. Modi, F. Bustreo, G. Conti, A. Malamitsi-Puchner, M. Hanson, Building resilient societies after COVID-19: the case for investing in maternal, neonatal, and child health. Lancet Public Health **5**, e624–e627 (2020). https://doi.org/10.1016/S2468-2667(20)30200-0

3. S. Whitelaw, M.A. Mamas, E. Topol, H.G.C. Van Spall, Applications of digital technology in COVID-19 pandemic planning and response. Lancet Digit. Health **2**, e435–e440 (2020). https://doi.org/10.1016/S2589-7500(20)30142-4

4. De', N. Pandey, A. Pal, Impact of digital surge during Covid-19 pandemic: a viewpoint on research and practice. Int. J. Inf. Manag. **55** (2020). https://doi.org/10.1016/j.ijinfomgt.2020.102171

5. R. Sharma, A. Shishodia, S. Kamble, A. Gunasekaran, A. Belhadi, Agriculture supply chain risks and COVID-19: mitigation strategies and implications for the practitioners. Int. J. Logist. Res. Appl. 1–27 (2020).https://doi.org/10.1080/13675567.2020.1830049

6. M.G. Garner, S.A. Hamilton, Principles of epidemiological modelling. OIE Rev. Sci. Tech. **30**, 407–416 (2011). https://doi.org/10.20506/rst.30.2.2045

7. C. Dube, G. Garner, M. Stevenson, R. Sanson, C. Estrada, P. Willeberg, The use of epidemiological models for the management of animal diseases. OIE 13–23 (2007)

8. K. Dodds, V.C. Broto, K. Detterbeck, M. Jones, V. Mamadouh, M. Ramutsindela, M. Varsanyi, D. Wachsmuth, C.Y. Woon, The COVID-19 pandemic: territorial, political and governance dimensions of the crisis. Territ. Polit. Gov. **8**, 289–298 (2020). https://doi.org/10.1080/21622671.2020.1771022

9. N.J. Brown, A novel peer-to-peer contact tracking application for COVID-19 and future pandemics. Diabetes Metab. Syndr. Clin. Res. Rev. **14**, 773–774 (2020). https://doi.org/10.1016/j.dsx.2020.06.001

10. M. Ghita, B. Zineb, B. Siham, C. Vi, W.G. Vi, Smart pandemic management through a smart, resilient and flexible decision-making system. **XLIV**, 7–8 (2020)

11. WHO (OMS), COVID-19 weekly epidemiological update. **1**, 4 (2020)

12. PRB, Population trends and challenges in the middle east and North Africa. MENA Policy Br. 1–8 (2001)

13. C. Connolly, S.H. Ali, R. Keil, On the relationships between COVID-19 and extended urbanization. Dialogues Hum. Geogr. **10**, 213–216 (2020). https://doi.org/10.1177/2043820620934209

14. Z.A. Nada Lebkiri, S. Jadda, A.S. Abdelrhani Mokhtari, Impact of containment type on Covid-19 propagation in Morocco using the SIR model. Bangladesh J. Med. Sci. **19**(Special: Introduction), (2020). https://doi.org/10.3329/bjms.v19i0.48167

15. ILO, With self-assessment checklist for RMG COVID-19 management guidance. **23** (2020)

16. O. Ifguis, M. El Ghozlani, F. Ammou, A. Moutcine, Z. Abdellah, Simulation of the final size of the evolution curve of coronavirus epidemic in Morocco using the SIR model. J. Environ. Public Health **2020**2020). https://doi.org/10.1155/2020/9769267

17. M. Serhani, H. Labbardi, Mathematical modeling of COVID-19 spreading with asymptomatic infected and interacting peoples. J. Appl. Math. Comput. (2020). https://doi.org/10.1007/s12190-020-01421-9

18. A. Bouchnita, A. Jebrane, A multi-scale model quantifies the impact of limited movement of the population and mandatory wearing of face masks in containing the COVID-19 epidemic in Morocco. Math. Model. Nat. Phenom. **15**,(2020).https://doi.org/10.1051/mmnp/2020016

19. M.U.G. Kraemer, C.H. Yang, B. Gutierrez, C.H. Wu, B. Klein, D. M. Pigott, L. du Plessis, N.R. Faria, R. Li, W.P. Hanage, J.S. Brownstein, M. Layan, A. Vespignani, H. Tian, C. Dye, O.G. Pybus, S.V. Scarpino, The effect of human mobility and control measures on the COVID-19 epidemic in China. Science (80) **368**, 493–497 (2020). https://doi.org/10.1126/science.abb4218

20. K. Chen, M. Wang, C. Huang, P.L. Kinney, P.T. Anastas, Air pollution reduction and mortality benefit during the COVID-19 outbreak in China. Lancet Planet. Health **4**, e210–e212 (2020). https://doi.org/10.1016/S2542-5196(20)30107-8

21. J.D. Berman, K. Ebisu, Changes in U.S. air pollution during the COVID-19 pandemic. Sci. Total Environ. **739**, 139864 (2020). https://doi.org/10.1016/j.scitotenv.2020.139864

22. M.K. James, M. Kishore, S.W. Lee, Demographic and socioeconomic characteristics of COVID-19 patients treated in the emergency Department of a New York City Hospital. J. Community Health (2020). https://doi.org/10.1007/s10900-020-00937-2

23. M.D. Pinheiro, N.C. Luís, COVID-19 could leverage a sustainable built environment. Sustainability **12**, (2020).https://doi.org/10.3390/su12145863

24. Z. Firano, F.A. Fatine, The COVID-19: macroeconomics scenarii and role of containment in Morocco. One Health **10**, 100152 (2020). https://doi.org/10.1016/j.onehlt.2020.100152

25. I. Cooper, A. Mondal, C.G. Antonopoulos, A SIR model assumption for the spread of COVID-19 in different communities. Chaos Solitons Fractals **139**, 110057 (2020). https://doi.org/10. 1016/j.chaos.2020.110057

26. H. Ben Hassen, A. Elaoud, N. Ben Salah, A. Masmoudi, A SIR-Poisson model for COVID-19: evolution and transmission inference in the Maghreb central regions. Arab. J. Sci. Eng. (2020). https://doi.org/10.1007/s13369-020-04792-0

27. B.R. Craig, T. Phelan, J.-P. Siedlarek, J. Steinberg, Improving epidemic modeling with networks. Econ. Comment (Federal Reserv Bank Cleveland) 1–8 (2020). https://doi.org/10.26509/ frbc-ec-202023

28. Y. Alharbi, A. Alqahtani, O. Albalawi, M. Bakouri, Epidemio-logical modeling of COVID-19 in Saudi Arabia: Spread projec-tion, awareness, and impact of treatment. Appl. Sci. **10** (2020). https://doi.org/10.3390/app10175895

29. N. Picchiotti, M. Salvioli, E. Zanardini, F. Missale, COVID-19 pandemic: a mobility-dependent SEIR model with undetected cases in Italy, Europe and US. (2020)

30. J.M. Carcione, J.E. Santos, C. Bagaini, J. Ba, A simulation of a COVID-19 epidemic based on a deterministic SEIR model. Front. Public Health **8**, (2020).https://doi.org/10.3389/fpubh.2020.00230

31. N. Wu, X. Ben, B. Green, K. Rough, S. Venkatramanan, M. Marathe, P. Eastham, A. Sadilek, S. O'banion, Predicting onset of COVID-19 with mobility-augmented SEIR model. medRxiv 2020.07.27.20159996 (2020)

32. P. Teles, A time-dependent SEIR model to analyse the evolution of the SARS-CoV-2 epidemic outbreak in Portugal. (2020)

33. A. Das, A. Dhar, S. Goyal, A. Kundu, Covid-19: analysis of a modified SEIR model, a comparison of different intervention strategies and projections for India∗

34. SimCOVID: open-source simulation programs for the Covid-19 outbreak (2020)

35. S. Peng, S. Yingji, Beware of asymptomatic transmission: study on 2019-nCoV prevention and control measures based on extended SEIR model. IEEJ Trans. Power Energy **140**, NL1_1-NL1_1 (2020). https://doi.org/10.1541/ieejpes.140.nl1_1

36. H. Kang, K. Liu, X. Fu, Dynamics of an epidemic model with quarantine on scale-free networks. Phys. Lett. Sect. A Gen. at Solid State Phys. **381**, 3945–3951 (2017). https://doi.org/10.1016/ j.physleta.2017.09.040

37. O. Kounchev, G. Simeonov, Z. Kuncheva, The TVBG-SEIR spline model for analysis of COVID-19 spread, and a tool for prediction scenarios. 1–21 (2019)

38. S. Changruenngam, D.J. Bicout, C. Modchang, How the individ-ual human mobility spatio-temporally shapes the disease trans-mission dynamics. Sci. Rep. **10**, 1–13 (2020). https://doi.org/10. 1038/s41598-020-68230-9

39. A. Weiss, M. Jellingsø, M. Otto, A. Sommer, Spatial and temporal dynamics of SARS-CoV-2 in COVID-19 patients : a systematic review and meta-analysis. EBioMedicine (2020). https://doi.org/ 10.1016/j.ebiom.2020.102916

40. Á. Briz-redón, Á. Serrano-aroca, A spatio-temporal analysis for exploring the effect of temperature on COVID-19 early evolution in Spain. Sci. Total Environ. **728**, (2020). https://doi.org/10.1016/j. scitotenv.2020.138811

41. F. Aràndiga, A. Baeza, I. Cordero-carrión, R. Donat, M.C. Martí, P. Mulet, D.F. Yáñez, A spatial-temporal model for the evolution of the COVID-19 pandemic in Spain including mobility. (2020). https://doi.org/10.3390/math8101677

42. M.M. Dickson, F. Santi, Modelling and predicting the spatio-temporal spread of Coronavirus disease 2019. Lancet **2019**, (2019)

43. S. Ardabili, A.R. Varkonyi-koczy, Coronavirus disease (COVID-19) global prediction using hybrid artificial intelligence method of ANN trained with grey wolf optimizer. (2020)

44. A. Bouchnita, A. Jebrane, A hybrid multi-scale model of COVID-19 transmission dynamics to assess the potential of non-pharmaceutical interventions. Chaos Solitons Fractals **138**, 109941 (2020). https://doi.org/10.1016/j.chaos.2020.109941

45. J. Dignan, Smart cities in the time of climate change and Covid-19 need digital twins. IET Smart Cities **2**, 109–110 (2020). https://doi. org/10.1049/iet-smc.2020.0071

46. Z. Yang, Z. Zeng, K. Wang, S.S. Wong, W. Liang, M. Zanin, P. Liu, X. Cao, Z. Gao, Z. Mai, J. Liang, X. Liu, S. Li, Y. Li, F. Ye, W. Guan, Y. Yang, F. Li, S. Luo, Y. Xie, B. Liu, Z. Wang, S. Zhang, Y. Wang, N. Zhong, J. He, Modified SEIR and AI prediction of the epidemics trend of COVID-19 in China under public health interventions. J. Thorac. Dis. **12**, 165–174 (2020). https://doi.org/10.21037/jtd.2020.02.64

47. A.A. Malik, T. Masood, R. Kousar, Repurposing factories with robotics in the face of COVID-19. Sci. Robot. **5**, 17–22 (2020). https://doi.org/10.1126/scirobotics.abc2782

48. N. Science, C. Phenomena, S. Shastri, K. Singh, S. Kumar, P. Kour, V. Mansotra, Chaos, Solitons and Fractals time series forecasting of Covid-19 using deep learning models: India-USA comparative case study. Chaos Solitons Fractals Interdiscip. J. Non-linear Sci. Nonequilibrium Complex Phenom. **140**, 110227 (2020). https://doi.org/10.1016/j.chaos.2020.110227

49. L.A. Amar, A.A. Taha, M.Y. Mohamed, Prediction of the final size for COVID-19 epidemic using machine learning: a case study of Egypt. Infect. Dis. Model. **5**, 622–634 (2020). https://doi.org/ 10.1016/j.idm.2020.08.008

50. Z. Car, S.B. Šegota, N. An, I. Lorencin, V. Mrzljak, Modeling the spread of COVID-19 infection using a multilayer perceptron. **2020** (2020)

51. K. Gostic, L. McGough, E. Baskerville, S. Abbott, K. Joshi, C. Tedijanto, R. Kahn, R. Niehus, J. Hay, P. De Salazar, J. Hellewell, S. Meakin, J. Munday, N. Bosse, K. Sherratt, R. Thompson, L. White, J. Huisman, J. Scire, S. Bonhoeffer, T. Stadler, J. Wallinga, S. Funk, M. Lipsitch, S. Cobey, Practical considerations for measuring the effective reproductive number R_t. medRxiv Prepr. Serv. Health Sci. 1–21 (2020). https://doi.org/10.1101/2020.06.18. 20134858

52. K.E. Harper, C. Ganz, Digital twin architecture and standards. 0–12 (2019)

53. F. Laamarti, H.F. Badawi, Y. Ding, F. Arafsha, B. Hafidh, S.A. El, An ISO/IEEE 11073 standardized digital twin framework for health and well-being in smart cities. IEEE Access **8**, 105950–105961 (2020). https://doi.org/10.1109/ACCESS.2020.2999871

54. L.G.J.R. Clapper, Clarification of geospatial intelligence. Jpn. J. Behav. Ther. **7**, 43–44 (1982)

55. M. Ghita, B. Siham, M. Hicham, A.E.M. Abdelhafid, D. Laurent, Geospatial business intelligence and cloud services for context aware digital twins development, in *Proceedings—2020 IEEE International Conference of Moroccan Geomatics, MORGEO 2020* (2020), pp. 21–26. https://doi.org/10.1109/Morgeo49228. 2020.9121889

Remote Sensing and Artificiel Intelligence

Opportunities for Artificial Intelligence in Precision Agriculture Using Satellite Remote Sensing

Asmae Dakir, Fatimazahra Barramou, and Omar Bachir Alami

Abstract

Precision agriculture has benefited from the development of emerging technologies like Internet of Things (IoT), big data, and artificial intelligence (AI). The huge amount of high-resolution remotely sensed data, the development of frameworks, and machine learning (ML) algorithms have made the analysis of raw data more advanced and precise. Artificial intelligence had unlocked a new perspective to solve sophisticated challenges in agriculture. The goal of this paper is to present recent techniques, algorithms, and methodologies using artificial intelligence (AI) in precision agriculture (PA) using satellite remote sensing, and concern recent studies were conducted in the latest years 2019–2020. The accent was also pointed to the potential of AI in precision agriculture, the challenges, future needs, and trends in the field.

Keywords

Artificial intelligence • Precision agriculture • Satellite remote sensing • Machine learning

1 Introduction

In the context of demography growth in the last decades and the pressure that was applied to natural resources to respond to the need of humankind and the climate change that has affected natural resources in many regions in the world especially agricultural resources, it was a necessity to find solutions more adaptable to exploit cultivated lands in a sustainable way [27]. The development of new technologies allowed responding to these challenges and had made farming more intelligent and sufficient. With the emergence of geospatial technologies like big data, the Internet of Things (IoT), and artificial intelligence [1, 28] in the last decades, several solutions to fulfill the requirements of increasing demand for food were found.

Artificial intelligence techniques, including machine learning, have been used for various applications [12]: the automation of yield prediction [8, 29], crop stress detection, crop recognition, and growth [21], the distinction of crop characteristics in particular crop biomass and canopy structure, disease detection [10], plants health monitoring, weed detection, and phenotype classification [26]. These technologies and tools have permitted to monitor spatial variability among farms and large crop fields that negatively affect crop growth and yields.

Usage of remote sensing technologies for PA has also increased rapidly during the past few decades. The unprecedented availability of high-resolution (spatial, spectral, and temporal) satellite images has promoted the use of remote sensing in many PA applications, including crop monitoring [7], irrigation management [29], nutriment application, disease and pest management [25], and yield prediction [35].

The application of artificial intelligence in agriculture has attracted the interest of a huge number of researchers. Mekonnen et al. [16], Pathan [23], and Talaviya [31] have cited multiple artificial intelligence techniques used in precision agriculture, but more focused on aerial and handheld remote sensing. Miriyala and Sinha [18] were more interested in deep learning to estimate crop yield.

This study is more focused on satellite remote sensing as the development of spectral, temporal, and spatial resolution in recent years has produced precise data with high quality and resolution.

A. Dakir (✉) · F. Barramou · O. B. Alami
Geomatics Science Research team (SGEO), LaGeS Laboratory,
Hassania School of Public Works, Casablanca, Morocco

© The Author(s), under exclusive license to Springer Nature Switzerland AG 2022
F. Barramou et al. (eds.), *Geospatial Intelligence*, Advances in Science, Technology & Innovation,
https://doi.org/10.1007/978-3-030-80458-9_8

2 Precision Agriculture

As the development of new digital technologies has trained a revolution in all life sides, new technologies have radically changed agricultural management and practices [12]. With all the benefits of new technologies including efficiency and sustainability, the domain of agriculture had taken advantage of these evolutions containing the employment of remote sensing data, automation, deep data analysis to manage agriculture smartly [23] at a regular interval of time. The domain of agriculture had thus known the emergence of the concept of precision agriculture or again precision farming that has in a prior purpose to have more with less; in other words, improving the productivity of crops with the optimization of resources. PA deals with the variability in the repartition of crops in the field. The International Society of Precision Agriculture adopted the following definition of precision agriculture in 2019: 'Precision agriculture is a management strategy that gathers, processes and analyzes temporal, spatial and individual data and combines it with other information to support management decisions according to estimated variability for improved resource use efficiency, productivity, quality, profitability and sustainability of agricultural production.'

AI had provided the required information to build the right knowledge to conduct efficient decision-making by providing smart irrigation techniques, ensuring the right pesticide requirements, detection of diseases, allowing the improvement of productivity, the quality of the crop, and avoid risks [31]. Tools provided by PA to drive and support complex decision-making have then allowed moving to 'digital agriculture' [27].

Artificial intelligence has proved its utility in smart irrigation as well. With the challenges of climate change that caused drought in lots of regions in the world, emerged the necessity of irrigating more areas with the low consumption of water [11].

3 The Potential of Artificial Intelligence in Precision Agriculture

Several studies have studied the application of smart technologies in precise farming like the Internet of Things (IoT), by using proximal sensings like smart tractors equipped with GPS and sensors, drones, aerial, and handheld remote sensing [1, 16, 28].

Artificial intelligence can be used in training robots to do the labor of tending, harvesting, and maintaining farmland efficiently that usually requires a lot of human capital, time, and effort. AI in agriculture application is emerging in three

areas: robotics, soil and crop monitoring, and predictive analytics [16].

The development of spectral, temporal, and spatial resolution in recent years has bounced the limitation of cloud cover as most of the data from the satellite images have a predefined wavelength and allowed the utilization of satellite remote sensing in precision agriculture that was more dedicated to proximal sensing.

Thus, AI has allowed several advantages, from the side of improving soil fertility, monitoring of growth rate of crops, smart irrigation that provides the exact amount of water in the exact frequency needed, providing pesticide effectively by identifying the ideal method to destroy weed plants [29] as well as yield mapping [35]. AI has also permitted the study and advancement in phenotyping with the very high spatial resolution satellite images attending 0.4 m [26]. The precise information provided by AI techniques have allowed decision-makers to predict and put efficient policies to manage the agricultural sector and guarantee food security [8].

4 Artificial Intelligence Applied to Precision Agriculture

4.1 Machine Learning

The uses of machine learning (ML) to date have fallen into two basic categories which are widely applicable in the field of agriculture and generally in remote sensing; the first category use ML for its regression capabilities and the second category uses machine learning for its classification capabilities. The ML is a part division of artificial intelligence that has the object to 'learn' and adapt through experience; its object is to extract information from data automatically using statistical methods.

ML application in PA can be categorized as crop management, livestock management, water management, and soil management [36]. ML application in crop management deals with yield prediction, disease detection, weed detection, and phenotype classification. In crop monitoring, traditional machine learning techniques can be used for prediction. However, the ability to learn the optimal features in the data is limited.

Liakos et al. [14] demonstrate different ML models used for solving real-world problems. The most commonly used ML models are artificial neural networks (ANNs) deep learning (DL), support vector machine (SVM), decision trees (DT), Bayesian models (BM), ensemble learning (EL), and dimensionality reduction (DR). There are numerous ML techniques available based on these ML models (Fig. 1).

Fig. 1 Examples of machine learning algorithms

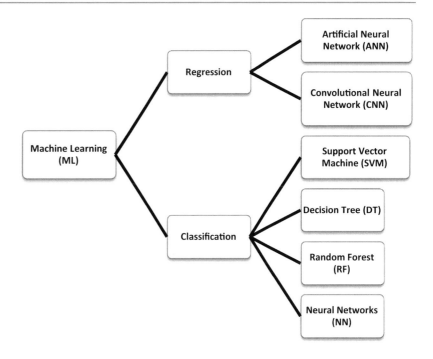

4.2 Classification Algorithms

Several classification algorithms based on machine learning (ML) are commonly used in agriculture especially crop mapping. The studies use object-learning algorithms or machine learning to identify crop types with optical and SAR datasets.

The support vector machine classifiers (SVM), decision tree (DT), and random forest (RF) have been the most common in recent years for the classification of images by remote sensing [3]. Many papers have reported better SVM performance [5, 15, 22] as well as DT and RF algorithms compared to other algorithms, including neural networks (NN). Despite the NN training phase can be time-consuming, requiring resources, and may require user experience, there are various advantages over support vector machine and decision tree algorithm, for instance, the rapid process of large volume images and probabilistic outputs of the classification [30]. In particular, RF is a non-parametric classifier based on classification and regression trees; it is used to generate supervised classification. Random forest has been implemented for various classification studies, by the reason of its robustness, fast, and simplicity [34].

For SAR images, maximum likelihood (ML), neural network classifiers (NN), random forest, and decision tree (DT) are the most common algorithms [30].

4.3 Regression Algorithms

- Artificial neural networks (ANN)

The artificial neural network (ANN) architecture is another algorithm in deep learning trying to deploy networks with good design for better generalization capability. The main advantage of using ANN is its ability to adapt to a changing environment, its robustness, and ability to learn itself by selecting appropriate values for weights [18]. This algorithm has been used recently in agricultural applications like the soil fertility prediction that was proposed by Song. Using ANN, the neural networks can handle the complex mapping using input variable sets [18]. Another application is the assessment of water stress by modeling thermal information [33].

- Convolutional neural networks (CNN)

Sinwar et al. [29] have exploited the convolutional neural network to predict diseases in crops by using captured images as a dataset for training. The algorithm extracts useful features from images and makes predictions. Sinwar concluded that the methodology was effective especially in continuous monitoring of the health of plants. This prediction allows therefore anticipating the ideal solutions to apply pesticides and the right time to use.

Wu et al. [36] have also exploited CNN-based machine learning methods and tree-based to predict soil properties with precision by producing geo-objects. The procedure is based on geo-object extracted from high-resolution remote sensing images. The method includes downscaling, raster aggregation, spatial overlay, machine learning model fitting, predictions, and data exportation. The CNN was mainly used to recognize geo-object with high edge precision from high-resolution images.

Among all deep learning methods, the convolution neural network has the highest impact on the performance of image classification and regression tasks [18].

4.4 Deep Learning

With the development of the resolution of satellite remote sensing, particularly the spectral and spatial resolution, emerged the necessity of robust methods and algorithms that handle and process that huge amount of data, in addition to the need to improve image analysis and classification and obtain maximum accuracy. Traditional ML techniques were not sufficient in producing the optimal decisions [29].

Deep learning is a class of machine learning algorithms that uses multiple layers to progressively extract higher-level features from the raw input. ResNet is an example of a residual learning framework forming a network of 152 layers to achieve efficient training in deep learning and solve the complex image classifications fast and flexibly. The next figure exposes the general structure of NN for predicting crop yield (Fig. 2).

4.5 Genetic Algorithms

A genetic algorithm is an optimization method based on the imitation of natural selection processes. The genetic algorithm uses a finite set of solutions, creating new data using the selection, mutation, and crossover operators. Output, external, and internal parameters are a quantitative assessment of the parameters of the object [9].

Several studies have used the convolution neural network and genetic algorithm in PA especially the search for diseases of crops. Korchagin et al. [9] have applied a genetic algorithm to solve the problem of diagnosing late plague on potato leaves and making forecasts [4]. They have exploited artificial intelligence by using both genetic algorithm and CNN applied to satellite remote sensing. The LAPAN-IPB images were first processed by CNN to classify plant types, then to allow the chromosome modeling. The genetic algorithm was then applied to give the final best solution. The study has demonstrated through that example that machine learning algorithms, especially genetic algorithms, and CNN can solve precision agriculture problems using satellite remote sensing.

5 Overall Review of AI Application in Precision Agriculture

Prediction and estimation of crop yields are among the major application that attracted the interest of researchers [1–3]. The developments of machine learning algorithms have shown great advantages over statistical methods that were used. Table 1 presents some studies that used MODIS imagery and neural network to estimate and predict crop

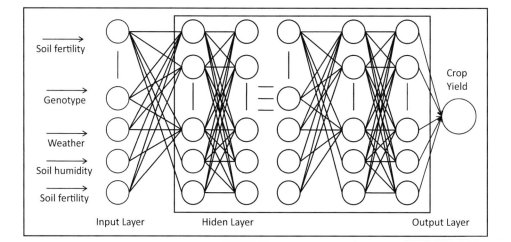

Fig. 2 General structure of neural network for the prediction of crop yield (redrawn based on the work of [6, 14])

Table 1 Applications of machine learning algorithms for crop yield estimation and prediction

Application	Crop type (and source)	Algorithm	Aim	Data imagery	Essential results
Crop yield estimation and prediction	Winter wheat [19]	The CNN network structure was designed for multitemporal MODIS images	Design a CNN network structure to extract features of winter wheat growth for yield estimation	19 estimation indexes were selected from 6 different MODIS products	The estimated yield of winter wheat based on time-series remote sensing images is highly correlated with statistical data, with Pearson's r of 0.82, RMSE of 724.72 kg hm^{-s2}
	Corn [8]	Deep neural networks with multiple hidden layers	Establishment of a deep neural network (DNN) approach to estimate crop yield	The dataset included 2,267 experimental hybrids planted in 2,247 of locations across the United States and Canada	The adopted method outperformed other popular methods such as Lasso, shallow neural networks (SNN), and regression tree (RT). The results also revealed that environmental factors had a greater effect on crop yield than genotype
	Winter wheat [32]	BP neural network and IPSO-BP neural network	Improving the estimation accuracy of regional winter wheat yields, and construct models using the weighted LAIs, the weighted VTCIs and I, respectively	Remotely sensed VTC and remotely sensed LAI extracted from MODIS LST data product	The regression models based on the IPSO-BP neural network were better than the regression models based on the BP neural network. There was a significant positive correlation between the estimation and the actual yields

yield. The challenge is to have precise predictions for better strategy and decision-making.

Crop disease studies are at the core of agricultural studies. The coupling of machine learning capabilities with remote sensing imagery has allowed detection, monitoring, and prediction of different crop diseases on a large scale, and thus taking sufficient decisions to apply interventions on a local scale.

Table 2 presents examples of crop diseases and methods applied for the analysis. Several ML algorithms were applied as NN and naïve Bayes, SVM and have shown their advantages over correlation methods [17].

One of the precise applications of ML in precision agriculture is the study of the phenotyping of crops (Table 3). With the concern of selecting the best variety of crops that gives the best throughput, better resistance to diseases, and stress-tolerant [37]; the intent then is to select the best varieties for the process of breeding giving improvement for the adaptability to conditions of climate change.

Table 2 Examples of applications related to crop diseases

Application	Crop type (and source)	Algorithm	Aim	Data imagery	Essential results
Detection, prediction, and evaluation of crop diseases	Coffee [17]	ML classification algorithms: the Python-based learning algorithm naïve Bayes, random forest, and the multilayer perceptron	Comparison of machine learning that can approximate the process of evaluating the disease in the field data	Landsat 8 OLI images	The best classifier performance was naïve Bayes and multilayer perceptron. Machine learning tools were more efficient than Pearson's correlation to detect the incidence of coffee necrosis
	Citrus [10]	(SVM), k-nearest neighbor (kNN), logistic regression (LR), naïve Bayes, neural network and ensemble learning	Investigating feature pre-processing and extraction, machine learning models and accuracy improvement on UAV multispectral images to explore infield detection of citrus HLB in a large area scale	Low-altitude remote sensing platform, and a multispectral camera (ADC-lite) was mounted on DJI M100 UAV to collect the data	Multispectral images with NIR, red, and green bands can effectively detect HLB under proper feature extraction and classification models
	Cassava [25]	The CNNs model	Proposing CNN algorithm to build a low-cost method to detect cassava infections through deep learning	10,000 labeled images collected during a regular survey	The model performance showed promising results for the classification of cassava mosaic and other cassava disease detection

Satellite remote sensing is promising in the field of phenotyping and phenology; with the availability of a higher spatial resolution, satellite images were more exploited to estimate crop traits. Several machine learning techniques were employed as decision tree (DT) and neural network. Regarding the need for a very high spatial resolution to characterize phenotyping; several studies used pan-sharpening to take advantage of the information from the panchromatic band.

With the emergence of precision agriculture, the concept of precision irrigation was also recently developed to respond to the challenge of better water management in a more critical environment. Smart irrigation systems were then established to better manage water resources and improve productivity [29]. Emergent techniques of AI and ML algorithms have also contributed to the development of precision irrigation with the employment of satellite remote sensing imagery.

Other applications include also land-use land-cover (LULC), expansion, and suitability of crops to have accurate monitoring of the spatial distribution of crops [13] (Table 4).

Table 3 Examples of applications related to biophysical crop parameter

Application	Crop type (and source)	Algorithm	Aim	Data imagery	Essential results
Biophysical crop parameters (Phenotype and phenology)	Dry bean [26] 'Unmanned aerial system and satellite-based high-resolution imagery for high-throughput phenotyping in dry bean'	Vegetation indices and correlations analysis was made with biomass and seed yield	To evaluate and establish remote sensing methods for high-throughput plant phenotyping in dry bean breeding nurseries	– Pleiades-1A (0.5 m) – SPOT 6 (1.5 m) – Planet scope (3.0 m) – Rapid Eye (5.0 m) + Aerial data	Sub-meter resolution satellite multispectral imagery showed promising application in field phenotyping, especially when the genotypic responses to stress are prominent
	Maize [6] 'Modeling maize above-ground biomass based on machine learning approaches using UAV remote-sensing data'	4 ML algorithms (MLR, SVM, ANN, and RF	– Comparing the performance of different machine learning modeling methods to estimate maize above ground biomass	Digital and multispectral imagery from unmanned aerial vehicle	– RF model gave the most balanced results – BIOVP can retain the largest strength effect on the AGB estimate in four different ML models by using importance analysis of predictors
	— [2] 'DATimeS: A machine learning time series GUI toolbox for gap-filling and vegetation phenology trends detection'	ML regression algorithms (MLRAs): DT, kernel-based methods, NN, and conventional fitting methods	The establishment of decomposition and analysis of time series software (DATimeS)	Satellite imagery (e.g.: Sentinel)	DATimeS aims to be a generic and innovative time-series toolbox that provides flexibility to capture the vegetation features, for gap-filling and subsequent phenological analysis

Table 4 Other precision agriculture applications

Application	Crop type (and sosurce)	Algorithm	Aim	Data imagery	Essential results
Evaluation of expansion and suitability/land-cover	Hickory [13]	Random forest (RF) approach/the maximum entropy-based (MaxEnt) model	Evaluation of expansion and suitability of hickory crop (land-cover map with emphasis on young and mature hickory plantations)	Chinese satellite Gaofen-1 images—Landsat imagery	The RF classifier based on multisource data can effectively produce a land-cover map with an overall accuracy of 89.5% and second, the MaxEnt model can be effectively used to evaluate suitability for hickory plantations
	20 land-cover and land-use classes [24]	A deep learning strategy for HSI classification that combines different convolutional neural networks	Classification map with 20 classes	Hyperspectral imagery (HSI)	Overall classification accuracy of 66.73% on the 2018 IEEE GRSS hyperspectral dataset—a high-resolution dataset that includes 20 land-cover and land-use classes
Precision irrigation and water stress	Maize [11]	Vegetation indices (NDVI, RECI, and RENDVI) were calculated then correlation and regression analysis (Pearson's correlation)	Determine the optimal time and depth of soil water content SWC and its relationship to maize grain yield	RapidEye Imagery	Superficial soil layers would be the most appropriate to monitor. Soil water content monitoring at reproductive stages combined with vegetation indices could be a tool for improving maize irrigation management
	— [25]	SVM—RF—eXtreme gradient boosting—Rotation forest—Oblique random forest—Artificial neural network	Overall review of the widely used methods for crop water stress monitoring using remote sensing and machine learning	Remote sensing data and measurements data	ANN and then SVM and RF were shown to be powerful for the classification and prediction of RS data

6 Opportunities, Challenges, and Future Trends

6.1 The Necessity of Expertise and Variability for Standardization of Treatments

Although the progress made by introducing artificial intelligence (AI) in precision agriculture (PA), there are multiple needs for future research in PA with remote sensing and cloud computing capabilities. First, considering the variability of general conditions (climate, soil, weather conditions, and crops), the methodologies and framework that were established remain specific to the conditions of application and cannot be reproducible in other areas. It is therefore recommended to understand the variability in space and time to reduce uncertainty in estimating crop parameters. For instance, Miriyala and Sinha [18] have

concluded that variability of the environment and genotype influence the accuracy of predictive models for crop yield estimation.

With the variability of remote sensing data and resources and multiplicity of methodologies and workflows, expertise is required to process data. It is then essential to develop a simple and reliable workflow for image pre-processing, analysis, and application in real time to facilitate the use of satellite data for the final customers.

6.2 The Complexity of Treatments and Costs

Recent works are presenting and modeling a single problem and the analysis includes its variables. The reality though presents different phenomenon that can interfere and influence each other. The needs embrace complex treatments that consider a more advanced algorithm for dimensionality reduction and time series smoothing [19]. For instance, AI has demonstrated its efficiency to monitor and detect crop diseases regardless of crop types. However, in the study of Korchagin et al. [9] the requirement is to analyze multiple factors, as the modelization of multiple co-occurring diseases on the same plant [25].

Other than the complexity of the treatment, there is also the challenge of the cost; especially for small farms, the installation of sensors and their maintenance [33], and the necessity of high spatial, spectral, and temporal resolution for remote sensing data can be unaffordable.

6.3 Data Availability

The availability of remote sensing archive images like IKONOS, LANDSAT, or SPOT should be exploited with real-time remote sensing images as well as other inputs like digital elevation models and soil maps to conduct precise analysis and model homogenous parts within the fields [20].

Current data publicly available are attending a spatial resolution of 0.31 m for WorldView-3 and 0.5 for the Pleiades that can be augmented by the Pansharpening treatment. The new improvements of spatial resolution satellite imagery are giving better perspectives, and for several applications they were sufficient as the estimation of crop yield, the evaluation of suitability, or monitoring of crop phenology. Though for some applications as throughput crop phenotyping [26], a further precise resolution is needed.

Spectral and temporal resolutions are also critical. For phenotyping studies, the definition of a specific wavelength in a given time can be unaffordable with rigid satellite remote sensing systems that have limited bands and defined periodicity [37]. New perspectives are yet to be coming with Maxar Technologies, Planet Labs, and Airbus that are planning new missions to improve spatial and temporal resolutions.

6.4 Algorithms

Multiple algorithms and methods were developed regarding each application and adapted to different variables. ML algorithms require an important amount of data for training to develop a more accurate model.

From multiple algorithms, particular ones were more adapted to agricultural applications as neural network (NN), support vector machine (SVM), and random forest (RF) [19, 33].

Machine learning algorithms coupled with remote sensing data have shown great potential in many agricultural applications. However, some applications were more employed like crop type mapping, disease monitoring, and yield prediction. The potential is yet to be explored for other applications that demand further precision like phenotyping applications [37]. Mu et al. [19] and Virnodkar et al. [33] have suggested the advantage of NN regarding other machine learning algorithms as SVM and RF, while Han et al. [6] have demonstrated that RF is more advantageous than SVM and ANN for estimating high-density biomass and gives more balanced results. Studies have also suggested the outperformance of deep neural networks that used multiple hidden layers over other current algorithms as a shallow neural network, regression tree, and Lasso [8].

7 Conclusion

Artificial intelligence (AI) had shown its efficiency in the different scientific fields as well as agricultural domain. This work presented a general overview of the principal applications of artificial intelligence in precision agriculture. Machine learning (ML) as a branch of AI that provides the system with the ability to learn from precedent experiences without being programmed had been widely used to respond to the specific challenges that have been imposed in a more demanding reality; beginning with a general prediction of crop yield and ending with more specific concern as capturing the phenotypic differences in the same field to maximize breeding efficiency. Machine learning models and algorithms have been widely deployed and developed for multiple agricultural needs (monitoring crop and phenological characteristics, smart irrigation, detection and prediction of crop diseases, etc.) using regression, classification methods, or again unsupervised algorithms. Studies have shown the advantages of using artificial neural network (ANN) and deep neural network (DNN) over other ML algorithms. The results can be more accurate with the usage of more amounts of data and archive information that allow the algorithms to be more accurate. ML techniques have demonstrated their efficiency to help in making better decisions to meet global food demand.

Perspective and future work

The employment of artificial intelligence, specifically machine learning algorithms, has demonstrated its advantages when coupled with satellite remote sensing data. A new perspective that was opened with new programs as Planet that gives very high spatial and temporal satellite imagery gives new opportunities to monitor agriculture. One of the applications that need precise analysis is the control irrigation of new plantations.

References

1. S. Bagwari, Impact of internet of things based monitoring and prediction system in precision agriculture. **22**, 4599 (2019)

2. S. Belda, L. Pipia, P. Morcillo-Pallarés, J.P. Rivera-Caicedo, E. Amin, C. De Grave, J. Verrelst, DATimeS: a machine learning time series GUI toolbox for gap-filling and vegetation phenology trends detection. Environ. Model. Softw. **127**, 104666 (2020). https://doi.org/10.1016/j.envsoft.2020.104666

3. A. Dakir, B. Omar, B. Fatimazahra, Crop type mapping using optical and radar images: a review (2020)

4. Firdaus, Y. Arkeman, A. Buono, I. Hermadi, Satellite image processing for precision agriculture and agroindustry using convolutional neural network and genetic algorithm. IOP Conf. Ser. Earth Environ. Sci. **54**, 012102. https://doi.org/10.1088/1755-1315/54/1/012102

5. J.K. Gilbertson, J. Kemp, A. van Niekerk, Effect of pan-sharpening multi-temporal Landsat 8 imagery for crop type differentiation using different classification techniques. Comput. Electron. Agric. **134**, 151–159 (2017). https://doi.org/10.1016/j.compag.2016.12.006

6. L. Han, G. Yang, H. Dai, B. Xu, H. Yang, H. Feng, Z. Li, X. Yang, Modeling maize above-ground biomass based on machine learning approaches using UAV remote-sensing data. Plant Methods **15**, 10 (2019). https://doi.org/10.1186/s13007-019-0394-z

7. X. Jiao, J.M. Kovacs, J. Shang, H. McNairn, D. Walters, B. Ma, X. Geng, Object-oriented crop mapping and monitoring using multi-temporal polarimetric RADARSAT-2 data. ISPRS J. Photogramm. Remote. Sens. **96**, 38–46 (2014). https://doi.org/10.1016/j.isprsjprs.2014.06.014

8. S. Khaki, L. Wang, Crop yield prediction using deep neural networks. Front. Plant Sci. **10** (2019). https://doi.org/10.3389/fpls.2019.00621

9. S. Korchagin, D. Serdechny, R. Kim, D. Terin, M. Bey, The use of machine learning methods in the diagnosis of diseases of crops. E3S Web Conf. **176**, 04011 (2020). https://doi.org/10.1051/e3sconf/202017604011

10. Y. Lan, Z. Huang, X. Deng, Z. Zhu, H. Huang, Z. Zheng, B. Lian, G. Zeng, Z. Tong, Comparison of machine learning methods for citrus greening detection on UAV multispectral images. Comput. Electron. Agric. **171**, 105234 (2020). https://doi.org/10.1016/j.compag.2020.105234

11. A. de Lara, L. Longchamps, R. Khosla, Soil water content and high-resolution imagery for precision irrigation: maize yield. Agronomy **9**, 174 (2019). https://doi.org/10.3390/agronomy9040174

12. D. Lary, Artificial intelligence in geoscience and remote sensing (2010)

13. G. Li, Z. Cheng, D. Lu, W. Lu, J. Huang, J. Zhi, S. Li, Examining hickory plantation expansion and evaluating suitability for it using multitemporal satellite imagery and ancillary data. Appl. Geogr. **109**, 102035 (2019). https://doi.org/10.1016/j.apgeog.2019.102035

14. K.G. Liakos, P. Busato, D. Moshou, S. Pearson, D. Bochtis, Machine learning in agriculture: a review. Sensors **18**, 2674 (2018). https://doi.org/10.3390/s18082674

15. U. Lussem, C. Hütt, G. Waldhoff, Combined analysis of SENTINEL-1 and Rapideye data for improved crop type classification: an early season approach for rapeseed and cereals. ISPRS Int. Arch. Photogram. Remote Sens. Spat. Inf. Sci. **XLI-B8**, 959–963 (2016). https://doi.org/10.5194/isprs-archives-XLI-B8-959-2016

16. Y. Mekonnen, S. Namuduri, L. Burton, A. Sarwat, S. Bhansali, Review—machine learning techniques in wireless sensor network based precision agriculture. J. Electrochem. Soc. **167**, 037522 (2020). https://doi.org/10.1149/2.0222003JES

17. J. da R. Miranda, M. de C. Alves, E.A. Pozza, H. Santos Neto, Detection of coffee berry necrosis by digital image processing of Landsat 8 Oli satellite imagery. Int. J. Appl. Earth Obs. Geoinf. **85**, 101983 (2020). https://doi.org/10.1016/j.jag.2019.101983

18. G. Miriyala, A. Sinha, Prediction of crop yield using deep learning techniques: a concise review (2019)

19. H. Mu, L. Zhou, X. Dang, B. Yuan, Winter wheat yield estimation from multitemporal remote sensing images based on convolutional neural networks (2019)

20. D.J. Mulla, Twenty five years of remote sensing in precision agriculture: key advances and remaining knowledge gaps. Biosys. Eng. **114**, 358–371 (2013). https://doi.org/10.1016/j.biosystemseng.2012.08.009

21. I. Neforawati, N.S. Herman, O. Mohd, Precision agriculture classification using convolutional neural networks for paddy growth level. J. Phys. Conf. Ser. **1193**, 012026 (2019). https://doi.org/10.1088/1742-6596/1193/1/012026

22. S. Park, J. Im, Classification of croplands through fusion of optical and SAR time series data. ISPRS Int. Arch. Photogramm. Remote Sens. Spat. Inf. Sci. **XLI-B7**, 703–704 (2016). https://doi.org/10.5194/isprs-archives-XLI-B7-703-2016

23. M. Pathan, N. Patel, H. Yagnik, M. Shah, Artificial cognition for applications in smart agriculture: a comprehensive review. Artif. Intell. Agric. **4**, 81–95 (2020). https://doi.org/10.1016/j.aiia.2020.06.001

24. K. Safari, S. Prasad, D. Labate, A multiscale deep learning approach for high-resolution hyperspectral image classification. IEEE Geosci. Remote Sens. Lett. 1–5 (2020). https://doi.org/10.1109/LGRS.2020.2966987

25. G. Sambasivam, G.D. Opiyo, A predictive machine learning application in agriculture: Cassava disease detection and classification with imbalanced dataset using convolutional neural networks. Egypt. Inform. J. (2020). https://doi.org/10.1016/j.eij.2020.02.007

26. S. Sankaran, J.J. Quirós, P.N. Miklas, Unmanned aerial system and satellite-based high resolution imagery for high-throughput phenotyping in dry bean. Comput. Electron. Agric. **165**, 104965 (2019). https://doi.org/10.1016/j.compag.2019.104965

27. M. Shepherd, J.A. Turner, B. Small, D. Wheeler, Priorities for science to overcome hurdles thwarting the full promise of 'digital agriculture' revolution. J. Sci. Food Agric. **100**, 5083–5092 (2020). https://doi.org/10.1002/jsfa.9346

28. R.K. Singh, M. Aernouts, M. De Meyer, M. Weyn, R. Berkvens, Leveraging LoRaWAN technology for precision agriculture in greenhouses. Sensors **20**, 1827 (2020). https://doi.org/10.3390/s20071827

29. D. Sinwar, V. Dhaka, M.K. Sharma, G. Rani, AI-based yield prediction and smart irrigation (2019), pp 155–180

30. S. Skakun, N. Kussul, A.Y. Shelestov, M. Lavreniuk, O. Kussul, Efficiency assessment of multitemporal C-band Radarsat-2 intensity and Landsat-8 surface reflectance satellite imagery for crop classification in Ukraine. IEEE J. Sel. Top. Appl. Earth Obs. Remote Sens. **9**, 3712–3719 (2016). https://doi.org/10.1109/JSTARS.2015.2454297

31. T. Talaviya, D. Shah, N. Patel, H. Yagnik, M. Shah, Implementation of artificial intelligence in agriculture for optimisation of irrigation and application of pesticides and herbicides. Artif. Intell. Agric. **4**, 58–73 (2020). https://doi.org/10.1016/j.aiia.2020.04.002

32. H. Tian, P. Wang, K. Tansey, S. Zhang, J. Zhang, H. Li, An IPSO-BP neural network for estimating wheat yield using two remotely sensed variables in the Guanzhong Plain, PR China. Comput. Electron. Agric. **169**, 105180 (2020). https://doi.org/10.1016/j.compag.2019.105180

33. S.S. Virnodkar, V.K. Pachghare, V.C. Patil, S.K. Jha, Remote sensing and machine learning for crop water stress determination in various crops: a critical review. Precis. Agric. **21**, 1121–1155 (2020). https://doi.org/10.1007/s11119-020-09711-9

34. F. Vuolo, M. Neuwirth, M. Immitzer, C. Atzberger, W.-T. Ng, How much does multi-temporal Sentinel-2 data improve crop type classification? Int. J. Appl. Earth Obs. Geoinf. **72**, 122–130 (2018). https://doi.org/10.1016/j.jag.2018.06.007

35. M.C.F. Wei, L.F. Maldaner, P.M.N. Ottoni, J.P. Molin, Carrot yield mapping: a precision agriculture approach based on machine learning. AI **1**, 229–241 (2020). https://doi.org/10.3390/ai1020015

36. T. Wu, J. Luo, W. Dong, Y. Sun, L. Xia, X. Zhang, Geo-object-based soil organic matter mapping using machine learning algorithms with multi-source geo-spatial data. IEEE J. Sel. Top. Appl. Earth Obs. Remote Sens. **12**, 1091–1106 (2019). https://doi.org/10.1109/JSTARS.2019.2902375

37. C. Zhang, A. Marzougui, S. Sankaran, High-resolution satellite imagery applications in crop phenotyping: an overview. Comput. Electron. Agric. **175**, 105584 (2020). https://doi.org/10.1016/j.compag.2020.105584

Monitoring Land Productivity Trends in Souss-Massa Region Using Landsat Time Series Data to Support SDG Target 15.3

Saadani Moussa, El Hassan El Brirchi, and Omar Bachir Alami

Abstract

The first step towards achieving a land degradation-neutral world and restoring degraded land and soil is to efficiently and accurately identify these lands at national and subnational levels. This step represents one of the targets within the Sustainable Development Goals (SDGs). It is worth recalling that the United Nations Convention to Combat Desertification (UNCCD) has adopted three sub-indicators for monitoring and assessing land degradation (Trends in Land Cover, Land Productivity, and Carbon Stocks). Through this study, we tried to take an important step in measuring the proportion of degraded land in the Souss-Massa region; we evaluated the sub-indicator Land Productivity at the pixel level (30 m) using Google Earth Engine (GEE). In addition, it should be emphasized that we used two decades of Landsat imagery and we chose the period between 2000 and 2015 as the baseline, while the comparison period was the one between 2016 and 2019. The results of this research showed that in the 4 years between 2016 and 2019, the Souss-Massa region has experienced remarkable changes in land degradation. We noted that 7.11% of the area of the region shows some improvement compared to 4.51% of the land that has degraded, while the rest of the zone which represents 88.38%, has not undergone any significant changes. To sum up, this study will constitute an important step in the assessment of land degradation in the Souss-Massa region and subsequently on all the Moroccan territory.

Keywords

Land degradation • Sustainable Development Goals (SDGs) • Landsat • Google earth engine • Land productivity • NDVI

S. Moussa (✉) · E. H. El Brirchi · O. B. Alami
LaGeS Laboratory, Geomatics Science Research Team 'SGEO', Hassania School of Public Works, Casablanca, Morocco

1 Introduction

At the beginning of the century, an estimated 2.6 billion people in more than 100 countries were affected by land loss and desertification, and more than 33% of the land area was influenced by this phenomenon [1]. For instance, Morocco is one of the most affected countries by Land Degradation.

The ability to effectively and accurately identify degraded land at different scales, from the local to the national level, will help to report progress towards Sustainable Development Goals (SDGs), particularly, the SDG Indicator 15.3.1 "Proportion of land that is degraded over total land area."

However, it is difficult to capture the status or condition of the land absolutely using a single indicator. Land Degradation is assessed and quantified according to three sub-indicators; Trends in land cover and Land Productivity can capture relatively rapid changes, while trends in above- and below-ground carbon stocks represent slower changes that indicate a trajectory or threshold approach.

These three parameters have a good accuracy and together they will assess the quantity and efficiency of terrestrial natural resources and the most related ecosystem services [2].

In this study, we focus on one of the sub-indicators, namely, Land Productivity, which we try to identify by presenting a new approach based on three main parameters; trend, which aims to assess the trajectory of primary productivity change over time; state, which is used to compare the productivity level of a given period with the productivity of the reference period; and performance, which refers to the efficiency of a given area relative to another area with similar productivity potential in the study area during the evaluation period.

In this research, two decades (2000–2019) of Landsat surface reflectance data were used from three sensors; Landsat 5, Landsat 7, and Landsat 8. It should be noted that these data have been corrected for atmospheric, reflectance effects, and satellite sensor discrepancies, with resolution of

© The Author(s), under exclusive license to Springer Nature Switzerland AG 2022
F. Barramou et al. (eds.), *Geospatial Intelligence*, Advances in Science, Technology & Innovation,
https://doi.org/10.1007/978-3-030-80458-9_9

(30 m), as well as the area of our study is the region of Souss-Massa, Morocco. Noting that is agreed that Remote Sensing and Earth Observation provide a potential tool to measure and identify areas where surface properties are changing due to land degradation [3].

We used Google Earth Engine (GEE); it is a cloud-based platform for rapid access to high-performance computing resources for the analysis of very large geospatial datasets [4].

2 Study Area

The Souss-Massa region is one of the 12 regions of Morocco Fig. 1, it covers over an area of 53,789 km^2, which represents 7.6% of the national territory. It occupies a stripe in the center of the country, stretching from the Atlantic Ocean to the western borders of Algeria. Moreover, it is a gateway between the Kingdom's North and South, enabling it to play a strategic role in the economic and socio-cultural levels. It is bounded to the north by the Marrakech-Safi region, to

the south by the Guelmim-Oued Noun region, to the east by the Drâa-Tafilelt region and Algeria, and to the west by the Atlantic Ocean.

Three factors determine the semi-arid climate of the region, namely, the relief, the oceanic coast, and the Sahara. Thus, the north of the region, dominated by the Atlas Mountains, is characterized by a humid to semi-arid climate as it progresses towards the plain. This latter, which occupies the lower relief of the Atlas Mountains as well as the basins of the Souss-Massa wadis, has an arid climate despite a wide opening on the Atlantic. Last, the southern and southeastern part of the region that makes up the northern side of the Sahara is covered by a desert climate.

Rainfall in the Souss plain has averaged 250 mm over the last 10 years, and 350–400 mm on the high plateaus. The southern part of the region, bordering the Sahara, is much drier, but since 2005, the desert has tended to green up, thanks to heavy winter rains, particularly in 2009–2010.

Besides, the winds are either east with desert influence, or west with ocean freshness [5].

Fig. 1 Map of Morocco highlighting Souss-Massa region

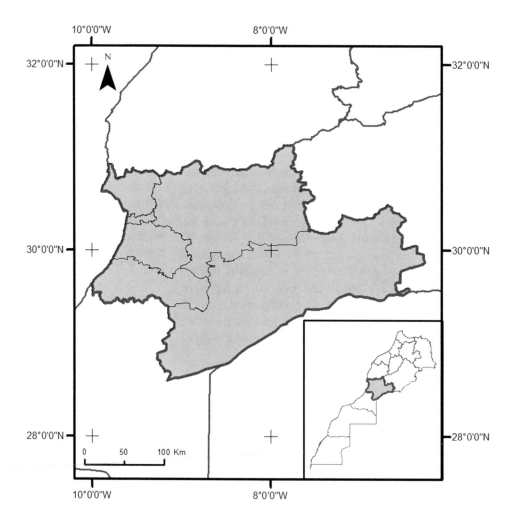

3 Data and Methods

3.1 Landsat Time Series and Composite Data

Two decades (2000–2019) of Landsat surface reflectance data were used in this research, which have been corrected for atmospheric, reflectance effects, and satellite sensor discrepancies.

3.2 Calculate Sensor Calibration Coefficients

In order to generate 20-year Landsat NDVI time series with high frequency, we used data from three sensors; Landsat 5, Landsat 7, and Landsat 8. Furthermore, we created a composite of images from this data.

As reported by numerous authors (Li and al. 2014) [6], (Roya and al. 2016) [7], and (Junchang and Masek 2017) [8], due to various spectral response functions between sensors, we have a small discrepancy in NDVIs.

Even though there are limited sensor discrepancies which can influence the evaluation of time series, we followed the methodology established by (Junchang and Masek 2017) to overcome this problem, based on the calibration of the cross-sensor and calculating the multiplication factor to produce the equivalent, the coefficient 1.036 adjusts Landsat 5 to its Landsat 7 equivalent, and the coefficient 1.086 adjusts Landsat 8 to its Landsat 7 equivalent.

3.3 Correction of Errors Due to Clouds, Cloud Shadows, and Haze Present on the Images

Clouds and their shadows can cause difficulties to optical sensors which can lead to errors in the detected trend. Several functions have been written in order to identify and classify any water, cloud, or apparent cloud shadow present on the images. These pixels were subsequently removed from the analysis. These functions were based on several quality assurance bands of the highest-level Landsat data product. Such a methodology has been used by (Braaten 2018) [9] and (Junchang and Masek 2017).

The quality assurance bands used for each sensor are illustrated in Table 1.

Table 1 Landsat sensor specific quality assurance bands

Sensor	Quality assurance bands
Landsat 5	pixel_qa, atmos_qa
Landsat 7	pixel_qa, atmos_qa
Landsat 8	pixel_qa, sr_aerosol

3.4 Land Cover Data

One of the essential aspects linked to monitoring land degradation is the definition of degradation in terms of changes in land cover so as to stratify and integrate them with other indicators, notably land productivity.

Land cover classification will serve as a reference for a more in-depth analysis and discussion of our study area. In order to assess productivity changes, we first worked on the calculation of land cover changes between 2000 and 2010.

The classification of land cover for 2000 is presented in Fig. 2 as well as the classification of land cover for 2019 is presented in Fig. 3:

3.5 Methodology

Land Productivity expresses the biological productive capacity of the land, as it is the main source of most human needs, whether for food, fiber, or fuel (United Nations Statistical Commission, 2016). Therefore, monitoring the productivity of the land and making every effort to maintain the high productivity of the land is one of the most important requirements for achieving sustainable development.

According to the metadata for SDG Indicator 15.3.1 [10], the changes in Land Productivity are among the determinants of the amount of land degradation. For example, the method of (Bai and al. 2008) [11] uses land productivity trends to cartograph land degradation using coarse resolution image data and adjusted climate effects by analyzing Rainfall Use Efficiency (RUE). However, this method is not suitable for our study because our study area is characterized by a semi-arid climate, whereas this method is specifically designed for areas with high rainfall, which makes it less suitable for areas with low vegetation cover (Wessels 2009) [12], as is the case for the Souss-Massa region.

In this study, we applied the methodology proposed by the Commonwealth Scientific and Industrial Research Organization (CSIRO) for the United Nations Convention to Combat Desertification (UNCCD). According to this method, land productivity is assessed by combining three metrics:

- Trend, which aims to assess the trajectory of primary productivity change over time,
- State, is used to compare the productivity level of a given period with the productivity of the reference period,
- Performance, which refers to the efficiency of a given area relative to another area with similar productivity potential in the study area during the evaluation period.

Fig. 2 Maps of land cover
(LC) 2000 for Souss-Massa
region

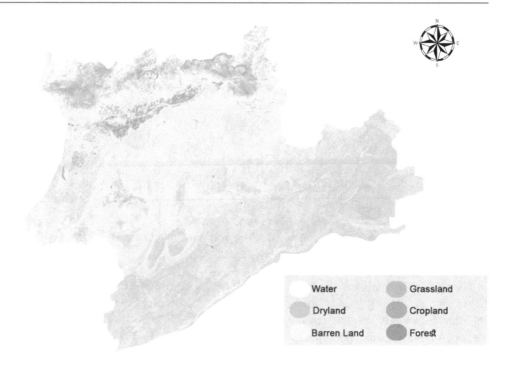

Fig. 3 Maps of land cover
(LC) 2019 for Souss-Massa
region

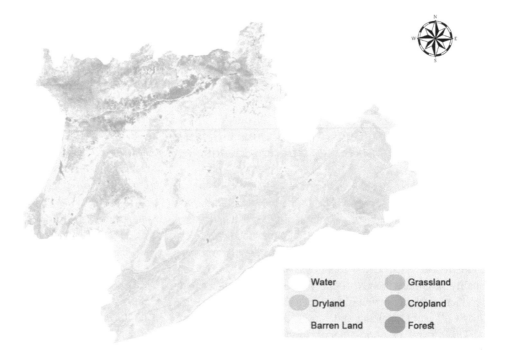

Net Primary Productivity (NPP) is the main variable used to monitor land productivity. There are multiple vegetation indices that can be derived from remote sensing data, and are closely related to NPP. However, the Normalized Difference Vegetation Index (NDVI) is among the most commonly used indices for NPP measurements.

The NDVI is produced from each scene's near infrared and red bands, as:

$$NDVI = \frac{NIR - Red}{NIR + Red} \qquad (1)$$

and its value ranges from −1 to 1.

The purpose of this study is to monitor land productivity in order to contribute to the reporting of the SDG Indicator 15.3.1. Hence, the metadata of this indicator has been taken as a reference. According to this metadata, it is sufficient to know whether productivity is increasing, decreasing, or stable at a given period and area, and not to measure the magnitude of the productivity evolution. For this reason, the NDVI can be considered good enough to give reliable results.

3.6 Calculating Productivity Metrics

3.6.1 Trend

The trend of productivity is determined with the aim of describing the path of change in the productivity of the land over time. This trend was calculated based on the yearly NPP values by developing a strong non-parametric linear regression model, as an instance, the Theil-Sen robust estimator (Ivits and Cherlet 2016) [13]. We relied on the Mann–Kendall score Z in order to determine the trend as there is a correlation between the scores of Z and the changeover in productivity; if Z is negative, this indicates a decline in productivity, while positive scores indicate an increase in productivity; this is what was referred to in the study of (Onyutha and al. 2016) [14].

Productivity trend assessments were calculated basing on the average annual productivity between 2000 and 2019 that is equivalent to 20 values, only important improvements those that display a p-value ≤ 0.05 were considered, this is what was described in Fig. 4.

According to the metadata for indicator 15.3.1, it is necessary to determine whether productivity is stable, increasing, or decreasing over time. We have divided the Z score into three sections where:

- Z score < −1.96: decreasing trend,
- Z score > 1.96: increasing trend,
- Z score > −1.96 AND < 1.96: Stable.

3.6.2 State

The productivity state index allows a comparison of the relative productivity level of a given period with the productivity of the reference period, i.e., per spatial unit or pixel (Ivits and Cherlet 2016). It is worth noting that for the purposes of this study, the period between 2000 and 2015 was chosen as the reference period, while the comparison period is the one between 2016 and 2019, the adoption of a 4-year period enables to avoid the changes related to annual climate fluctuations.

The assessment of changes in the productivity state is as follows; firstly, we calculated the annual integrals of NDVI for all years of the reference period, then we added 5% at both ends of the distribution interval to avoid having extreme values in NDVI for the out-of-interval comparison period. Afterwards, we used the frequency distribution curve to classify the annual productivity estimates into 10 classes. Next, the average NDVI index was calculated for the baseline and comparison periods, then we set the class to which each pixel corresponds, where possible values vary between

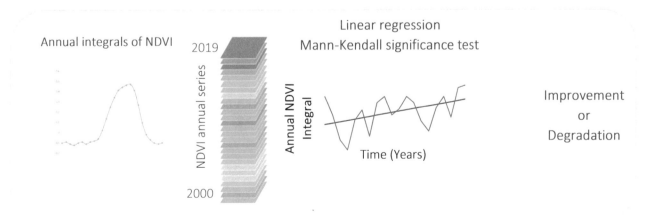

Fig. 4 The steps of productivity trend assessments

Fig. 5 Illustration of the assessment of changes in productivity state

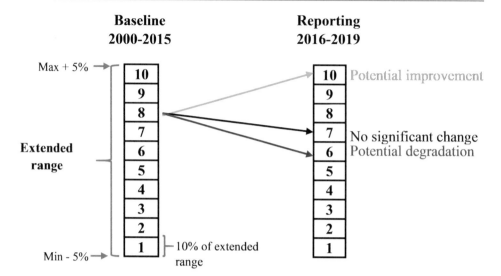

1 and 10. As shown in Fig. 5, for each pixel, we calculated the difference in class number between the reference period and the reporting period; if a pixel has a difference value less than or equal to −2, then it has a potential degradation, while if the difference value is greater than or equal to 2, then the pixel has a potential improvement, otherwise it is reported as stable.

3.6.3 Performance

Productivity performance refers to the efficiency of a given area relative to another area with similar productivity potential in the study area during the evaluation period. It is important to note that it is necessary to divide the Study Area into units of equal productive capacity based on factors such as phenology, moisture availability, and soil conditions (Ivits and Cherlet 2016) and (Ivits and al. 2013) [15]. In this study, we relied on a combination of soil classification units

provided by SoilGrid [16], which has recently become available on Google Earth Engine, and the land cover classification that we had prepared with an accuracy of 30 m.

In order to measure the Performance indicator, we initially classified the study area into ecologically identical units using the intersection of land cover and soil grid, then, we extracted all the mean NDVI values for each land unit and generated the frequency distribution. From this latter, the value representing the 90th percentile was derived, which is called the highest productivity for that unit. The ratio between mean NDVI (P_{obs}) and maximum productivity (P_{max}) represents the performance indicator.

$$\text{Performance} = \frac{P_{obs}}{P_{max}} \qquad (2)$$

The performance value which is less than 0.5 may indicate degradation as illustrated in Fig. 6.

Fig. 6 Illustration of the productivity performance measuring steps

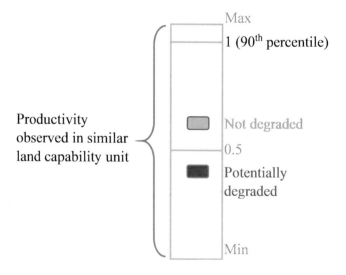

Table 2 Table showing the possible aggregation of productivity sub-indicators

Trend	State	Performance	Productivity
Improvement	Improvement	Stable	Improvement
Improvement	Improvement	Degradation	Improvement
Improvement	Stable	Stable	Improvement
Improvement	Stable	Degradation	Improvement
Improvement	Degradation	Stable	Improvement
Improvement	Degradation	Degradation	Degradation
Stable	Improvement	Stable	Stable
Stable	Improvement	Degradation	Stable
Stable	Stable	Stable	Stable
Stable	Stable	Degradation	Degradation
Stable	Degradation	Stable	Degradation
Stable	Degradation	Degradation	Degradation
Degradation	Improvement	Stable	Degradation
Degradation	Improvement	Degradation	Degradation
Degradation	Stable	Stable	Degradation
Degradation	Stable	Degradation	Degradation
Degradation	Degradation	Stable	Degradation
Degradation	Degradation	Degradation	Degradation

3.7 Aggregation of the Productivity Sub-indicators

The three sub-indicators of productivity (Trend, State, Performance) are then combined to indicate whether a pixel is degraded. They are combined as detailed in Table 2.

4 Results

Through the use of several sensors of Landsat data (Landsat 5, Landsat 7, and Landsat 8), we have provided a composite of images with 30 m resolution, and by following the methodology described in the part above, we have measured the three sub-indicators of land productivity (trend, state, and performance), as well as the logical matrix combination of these three sub-indicators which represents Land Productivity indicator.

4.1 Trend

Over the 20-year assessment period, we identified the trajectory of productivity trend for each pixel over our study area. From Table 3 and Fig. 7, it can be seen that 3.37% of the area had a positive productivity trend which indicates an improvement in the land condition of this part, while 2.65%

Table 3 Summary of change in productivity trend (2000 to 2019) for Souss-Massa region

Trend	Area (km^2)	% of total land area
Improvement	1812.68	3.37
Stable	50,550.90	93.98
Degradation	1425.40	2.65

of the area showed a negative productivity trend which is a sign of a degradation of land condition. The rest of the area did not register any significant change in productivity trend.

4.2 State

From the previously mentioned, the productivity state indicator allows discovering the recent evolutions of primary productivity in relation to the reference period which was defined in 2000–2015. The results showed that the state of productivity has been moderately changed, especially in the boundaries between drylands and grasslands.

From Table 4 and Fig. 8, it can be seen that 4.82% of the study area has registered a degradation in productivity state compared to 2.88% who have seen a recent improvement, the other area has seen small changes and is considered stable.

Fig. 7 Map of productivity trend (2000 to 2019) in Souss-Massa region

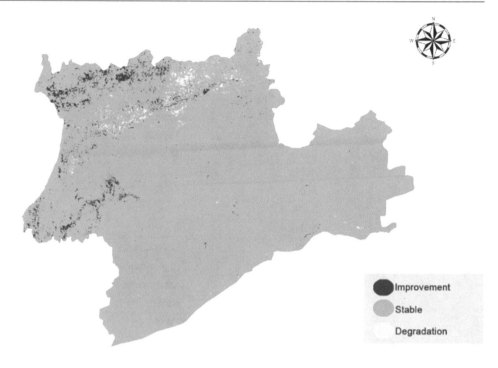

Table 4 Summary of change in Productivity State (2000–2019) for Souss-Massa region

	Area (km^2)	% of total land area
Improvement	1549.12	2.88
Stable	49,647.24	92.30
Degradation	2592.63	4.82

Fig. 8 Map of productivity state (2000 to 2019) in Souss-Massa region

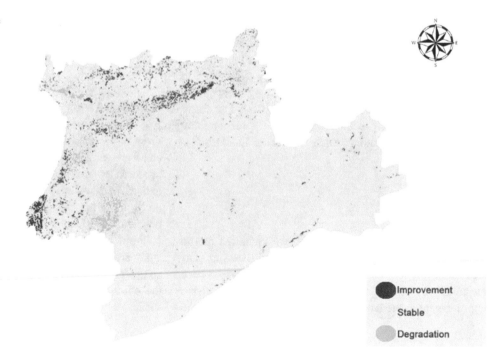

Fig. 9 Map of productivity performance (2000 to 2019) in Souss-Massa region

4.3 Performance

In order to measure Productivity Performance, we classified our area into homogeneous units based on a combination of soil grid and land cover classification, and subsequently identified productivity performance using the methodology cited above. As shown in Fig. 9 and Table 5, the productivity performance decreased in more 6% of the study area, and this degradation is especially marked in the grassland.

4.4 Combination of Productivity Indicators

The three productivity sub-indicators are ultimately combined into three classes which indicate the degradation of each pixel. Table 6 presents a summary of changes in land productivity in the Souss-Massa region. We noticed that 7.11% of the region shows an improvement compared to 4.51% of the land under study which has degraded, and the rest of the region which is 88.38% is stable.

Fig. 10 Map of land productivity (2000–2019) in Souss-Massa region

Table 5 Summary of change in productivity performance (2000–2019) for Souss-Massa region

	Area (km^2)	% of total land area
Stable	50,120.59	93.18
Degradation	3668.41	6.82

Table 6 Summary of changes in land productivity (2000–2019) for Souss-Massa region

	Area (km^2)	% of total land area
Improvement	3824.40	7.11
Stable	47,538.72	88.38
Degradation	2425.88	4.51

Table 7 Summary of change in land productivity sub-indicators (2000–2019) for Souss-Massa region

	Trend (%)	State	Performance	Productivity
Improvement	3.37	2.88	6.24	7.11
Stable	93.98	92.30	86.94	88.38
Degradation	2.65	4.82	6.82	4.51

5   Discussion and Conclusions

Through this study, we aimed to contribute to the evaluation of the SDG Indicator 15.3.1 "Proportion of land that is degraded over total land area," where we focused on the evaluation of land productivity according to the methodology proposed by the Commonwealth Scientific and Industrial Research Organization (CSIRO) for the United Nations Convention to Combat Desertification (UNCCD) in addition to the SDG Indicator 15.3.1 metadata. Our added value lies in the adoption of a data resolution of 30 m while most of the studies done in this perspective were with a resolution of at least 300 m; and this is the case for the Land Degradation Monitoring Toolbox Trends.Earth [17].

To accomplish this work, it was necessary to use thousands of satellites imagery, which would constitute a form of big data. Where it is impossible to process this huge amount of data by traditional means and methods. That's why, in this study, we used the cloud computing "Google Earth Engine," and with this engine, we were able to do the necessary work. It should be mentioned that the study area is very large, and despite the power of the GEE, we sometimes had to parallel the data processing operations or divide the study area into several parts by respecting the methodology.

Independently of the use of GEE technology, the creation of composite images emanating from three sensors (Landsat 5, Landsat 7, and Landsat 8) has permitted to provide a series of mages with a high temporal frequency, which has contributed to improve the quality of the results.

In conclusion, we have assessed the Land Productivity in the Souss-Massa region over an area of 53,789 km^2 using time series for Landsat data from 2000 to 2019 by adopting as reference the period 2000–2015. The results of this study will constitute an important step in the assessment of land degradation in the Souss-Massa region and subsequently on all the Moroccan territory. There is an urgent need to monitor changes in land productivity assessment with considerable accuracy, reliability, efficiency, and sustainability in order to increase the identification of land degradation for more effective and efficient resource management and land conservation. In our future work, we will focus on improving the mapping of land productivity and land use changes to identify degraded land, and also to combine the results of the study with social and economic data to determine the impact of land degradation on sustainable development.

References

1. C. Adams, H. Eswaran, *Global Land Resources in the Context of Food and Environmental Security. Advances in Land Resources Management for the 20th Century* (Soil Conservation Society of India, New Delhi, 2000), pp. 35–50.
2. B.J. Orr, A.L. Cowie, V.M. Castillo Sanchez, P. Chasek, N.D. Crossman, A. Erlewein, G. Louwagie, M. Maron, G.I. Metternicht, S. Minelli, A.E. Tengberg, S. Walter et S. Welton, *Scientific Conceptual Framework for Land Degradation Neutrality. A Report of the Science-Policy Interface* (Convention des Nations unies sur la lutte contre la désertification (CNULCD), Bonn, Allemagne, 2017). https://www.unccd.int/sites/default/files/documents/2019-06/LDN_CF_report_web-english.pdf
3. J.V. Vogt, U. Safriel, G. Von Maltitz, Y. Sokona, R. Zougmore, G. Bastin, J. Hill, Monitoring and assessment of land degradation and desertification: towards new conceptual and integrated approaches. Land Degrad. Dev. **22**(2), 150–165 (2011). https://doi.org/10.1002/ldr.1075
4. N. Gorelick, M. Hancher, M. Dixon, S. Ilyushchenko, D. Thau, R. Moore, Google earth engine: planetary-scale geospatial analysis

for everyone. Remote Sens. Environ. **202**, 18–27 (2017). https://doi.org/10.1016/j.rse.2017.06.031

5. Official website of the High Commission for Planning, Morocco Available at: https://www.hcp.ma/region-agadir/Presentation-de-la-region-de-Souss-Massa_a16.html

6. P. Li, L. Jiang, Z. Feng, Cross-comparison of vegetation indices derived from landsat-7 enhanced thematic mapper plus (ETM+) and landsat-8 operational land imager (OLI) sensors. Remote Sens. **6**, 310–329 (2014)

7. D.P. Roy, V. Kovalskyy, H.K. Zhang, E.F. Vermote, L. Yan, S.S. Kumar, A. Egorov, Characterization of Landsat-7 to Landsat-8 reflective wavelength and normalized difference vegetation index continuity. Remote Sens. Environ. **185**, 57–70 (2016). https://doi.org/10.1016/j.rse.2015.12.024

8. J. Ju, J.G. Masek, The vegetation greenness trend in Canada and US Alaska from 1984–2012 Landsat data. Remote Sens. Environ. **176**, 1–16 (2016). https://doi.org/10.1016/j.rse.2016.01.001

9. J. Braaten, *LandsatLinkr (LLR) Automated Landsat Image Processing System* (Oregon State University, Laboratory for Applications of Remote Sensing in Ecology, Corvallis, 2018)

10. Metadata for SDG Indicator 15.3.1: Proportion of land that is degraded over total land area: https://unstats.un.org/sdgs/metadata/files/Metadata-15-03-01.pdf

11. Z.G. Bai, D.L. Dent, L. Olsson, M.E. Schaepman, Proxy global assessment of land degradation. Soil Use Manag. **24**, 223–234 (2008). https://doi.org/10.1111/j.1475-2743.2008.00169.x

12. K. Wessels, Letter to the editor: Comments on 'Proxy global assessment of land degradation 'by Bai and al. (2008). Soil Use Manag. **25**, 91–92 (2009)

13. E. Ivits, M. Cherlet, Land-productivity dynamics towards integrated assessment of land degradation at globalscales title. EUR 26052. https://doi.org/10.2788/59315

14. C. Onyutha, H. Tabari, M.T. Taye, G.N. Nyandwaro, P. Willems, Analyses of rainfall trends in the Nile River Basin. J. Hydro-Environ. Res. **13**, 36–51 (2016). https://doi.org/10.1016/j.jher.2015.09.002

15. E. Ivits, M. Cherlet, W. Mehl, S. Sommer, Ecosystem functional units characterized by satellite observed phenology and productivity gradients: a case study for Europe. Ecol. Indicators **27**, 17–28 (2013). https://doi.org/10.1016/j.ecolind.2012.11.010

16. System for global digital soil mapping, available online at https://soilgrids.org/

17. Trends.Earth. Conservation International. Available online at: http://trends.earth

Subimages-Based Approach for Landslide Susceptibility Mapping Using Convolutional Neural Network

Mouad Alami Machichi, Abderrahim Saadane and Peter L. Guth

Abstract

Landslides are some of the deadliest and most violent geological events. A lot of research has been done on this topic in order to understand its causes and propose solutions. An essential tool for landslide risk management is landslide susceptibility maps. In this paper, we developed a Convolutional Neural Network (CNN) model capable of producing a susceptibility map using seven explanatory variables: lithology, slope, drainage density, fault density, elevation, roughness, and aspect. A susceptibility index map was generated in the Aknoul Region in the Rif to illustrate the CNN results. We found that areas with very high susceptibility index are affected the most by landslides.

Keywords

Landslide susceptibility • Deep learning • Convolutional neural networks • Geospatial modeling • Moroccan landslides

1 Introduction

A landslide can be defined as a slope failure, a mass movement of rocks, debris soil, or even organic material under the effect of gravity [1, 2]. Landforms resulting from this type of movement are also called landslides [3]. This geological phenomenon is among the deadliest natural disasters, causing a great deal of monetary damage and claiming each year the lives of thousands of people worldwide [3].

A lot of research has been done to characterize landslides and identify their varying causes [4, 5]. These studies have shown that landslides are the result of a combination of multiple conditioning factors intrinsic to the environment such as morphology, geology, and hydrology [6]. Areas that have been affected by the aforementioned factors have a very high probability of producing landslides if a triggering agent such as torrential rain or an earthquake were to happen [7, 8].

What makes this phenomenon difficult to study is the complexity of its factors and the interactions they have between each other. A myriad of methods has been developed for the purpose of Landslide Susceptibility Mapping (LMS). Most notably is the decision-based method AHP (analytic hierarchy process), logistic regression, bivariate regression, and, recently, Deep Learning (DL) models started to get used in LSM. DL were shown to outperform classic machine learning algorithms on multiple occasions [9, 10]. This is mainly because DL are more robust and more adapted in processing large and complex amounts of remote sensing data [11].

Input landslide data are crucial to the success of the LSM. Regardless of the LSM approach, we found that in most, if not all, of the studies, landslides were represented by points (centroids). This representation is reductionist and leads to a loss of very important information about the landslide genesis, even when neighboring pixels are added to create multidimensional subimages. Hence, the decision to represent landslides with (3×3) grids located around the trigger area, located at the top of each landslide in order to get the most accurate representation.

Our goal in this paper is to develop a convolutional neural network model that is capable of producing the landslide susceptibility model of the Rif (North of Morocco) (Fig. 1) using remote sensing data and open-source geospatial and deep learning tools.

M. A. Machichi (✉)
Moroccan Foundation for Advanced Science, Innovation & Research, Rabat, Morocco

A. Saadane
Department of Geology, Faculty of Sciences of Rabat, University Mohammed V, Rabat, Morocco

P. L. Guth
US Naval Academy, Annapolis, MD, USA
e-mail: pguth@usna.edu

© The Author(s), under exclusive license to Springer Nature Switzerland AG 2022
F. Barramou et al. (eds.), *Geospatial Intelligence*, Advances in Science, Technology & Innovation,
https://doi.org/10.1007/978-3-030-80458-9_10

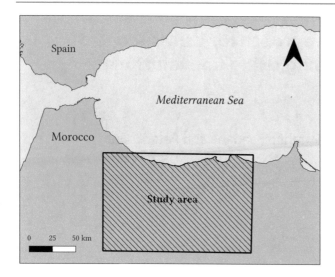

Fig. 1 Study area

2 Materials and Methods

2.1 Data

Most of the parameters that were used in this study were derived from the following geomorphological and geological data.

Geological maps

Geological maps are essential for generating ground truth data, as well as lithology and fault maps. In this study, ten 1:50000 geological maps were used: Al Hoceima [12], Midar [13], Rouadi [14], Boudinar [15], Ain Zohra [16], Aknoul [17], Taineste [18], Beni Ahmed [19], Ban El Mrouj-Taza [20], and Ain Bou Kellal-Msoun [21].

Landslide database

To create the landslide ground truth database, 435 landslides were manually digitized from the aforementioned geological maps of the Rif area in the north of Morocco as a reference.

Digital Elevation Models

Digital Elevation Models (DEM) are of great importance for landslide susceptibility mapping. Five out of the seven (elevation, slope, aspect, roughness, and the drainage map) conditioning factors used in this study were derived from DEMs. In this study, we used Tandem-X with a spacing of 0.4 arc second [22] that was provided by the German Aerospace Center.

2.2 Geospatial Modeling Tools

QGIS

QGIS is a free and open-source geographic information system (GIS) software that allows for the visualization, editing,

and analysis of geospatial data [23]. QGIS was used to geo-reference and digitize geological maps, preprocess digital elevation models, and generate some conditioning factors.

Keras

Keras [24] is a high-level API (applicative programming interface) that provides a convenient way to create almost any kind of deep learning model. In this study, we used Keras to develop, train, and test the deep learning model used.

2.3 Conditioning Factors

Landslides are some of the most complex geological hazards. Researchers have found that they are the results of a combination of causes. These causes are categorized into two broad categories: **conditioning factors** that weaken the slopes and make them unstable over a relatively long period; the second category of factors is called **trigger factors**, unlike the first category, these causes occur over a relatively short period and are unpredictable for the most part.

In this study, we used seven landslide conditioning factors as indicators for susceptible areas: lithology, slope, drainage density, fault density, hypsometry, roughness, and aspect. These factors are either directly recommended, or are derivatives of the landslide conditioning parameters recommended in [25].

Lithology

The type and competence of rock formations are important parameters that inform about the slope resistance and how easily it could fail. Quartzite, for example, is very resistant to erosion, whereas slopes that are mostly made of friable rock formations such as marl and flysch are easily alterable.

It is well known that the destabilization of slopes is conditioned and facilitated by the friable material, especially in areas where such material is abundant [5].

Figure 2 shows the digitized lithology of the Aknoul Region in the Rif.

Slope

The slope (Fig. 3) degree plays an important role in shaping the morphology of hillslopes. It influences the hydrology of a watershed and therefore the various instabilities that might occur inside it [6]. The stagnant water puddles, for example, that were used in [26] as indicator to detect landslides were the result of counter-slopes.

Drainage density

The erosive power of water streams can lead to the destabilization of slopes. Several reported landslides are reactivated along some of those streams especially during periods of flooding [27].

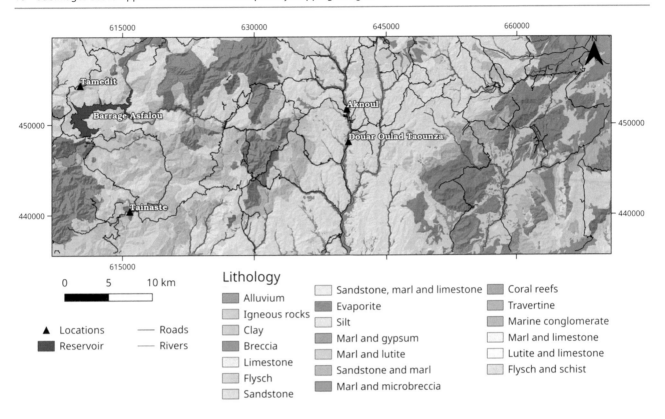

Fig. 2 Lithology map of the Aknoul Region

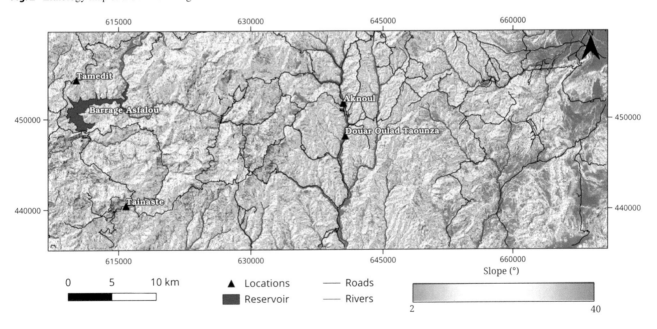

Fig. 3 Slope map of the Aknoul Region

Figure 4 shows the drainage density of the Aknoul Region in the Rif.

Fault density

Fractured rock formations present discontinuities that allow for different forms of instability. The faults were digitized

from the 10 geological maps, the resulting polyline was used to calculate the density of the faults in the study area (Fig. 5).

Hypsometry

Hypsometry (Fig. 6) is one of the most important landslide conditioning parameters [27]. In fact, this parameter has a

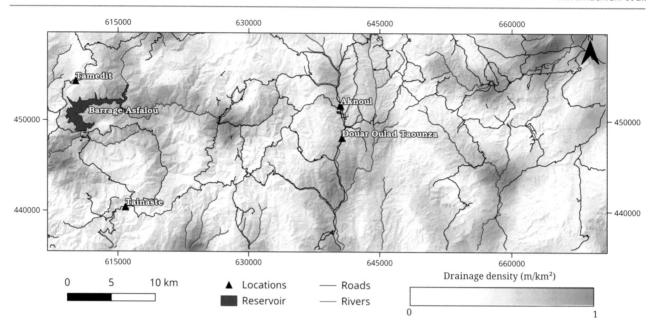

Fig. 4 Drainage density map of the Aknoul Region

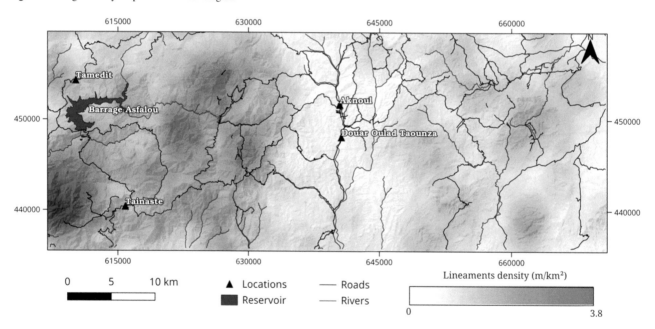

Fig. 5 Fault density map of the Aknoul Region

strong correlation with rainfall, which is one of the triggers of landslides.

Roughness

Roughness (Fig. 7) is defined as a topographic irregularity in a given surface [28]. Several studies [29–31] have used this parameter to delineate and map existing landslides.

Aspect

Slope aspect (Fig. 8) is a parameter that plays an important role in influencing the slope stability, because the type and rate at which rock formations are altered is directly related to the aspect. Moreover, soil development as well as vegetation type and growth are controlled by the slope aspect.

2.4 Deep Learning

Deep Learning (DL) is a relatively new sub-field of Machine Learning (ML) [32]. It has completely changed the state of the art in many fields such as computer vision, image processing, and speech recognition.

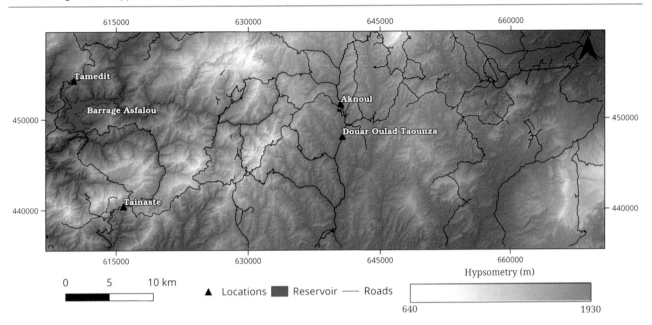

Fig. 6   Elevation map of the Aknoul Region

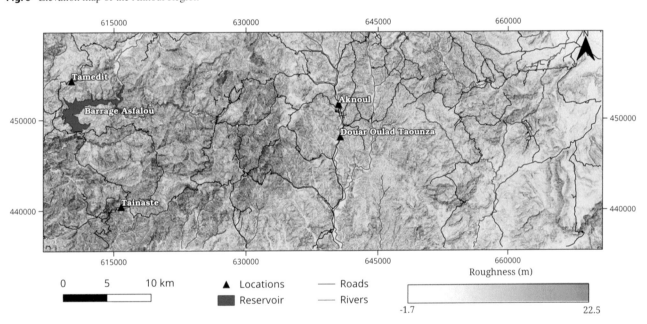

Fig. 7   Roughness map of the Aknoul Region

DL draws its name from the number of successive hidden layers that it is made of (depth). It is in contrast to the *Shallow Learning* (another name for ML) where only one or two layers of data representations are used [33].

Convolutional Neural Networks

CNN is a type of neural network designed to process data that come in the form of multiple arrays [32]. This type of data is very common for signals (1D arrays) and images (2D arrays). In the case of satellite imagery, each band can be represented by a single 2D array.

What makes CNN powerful is their ability to recognize patterns that they learn anywhere. Hence, because the patterns they learn are translation invariant, they need less training samples than regular dense layers in order to make generalizations [33]. Moreover, they can learn spatial hierarchies of patters, and they can extract increasingly complex and abstract visual concepts of the input data.

Input data

To create the input data for the model, a grid that covers the extent of the study area was generated with a horizontal and

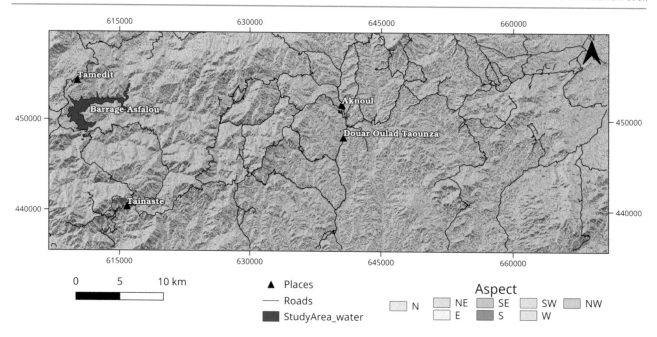

Fig. 8 Aspect map of the Aknoul Region

vertical spacing of 30 m. 20,800 subimages were then annotated with the following criteria: a value of 1 if the subimage
is located around the trigger area of a landslide, and a value of
0 otherwise. The conditioning factors were then stacked and
cropped using the created grids.

Numerical parameters had to be rescaled between 0 and
1 to ensure the homogeneity of the data. In this study, we
had one categorical parameter (lithology). This parameter was
converted into a numerical one before being inputted into the
model. There are a lot of techniques that are used to preprocess
categorical data. The method used in this study is called one-
hot encoding.

This technique creates a binary array of size N, where N
is the number of classes.

Architecture

The model used in this study has one input layer, four hidden
layers, and an output layer.

The goal behind using these layers is to create meaningful
representations of the input data [33].

Hidden layers are made of units called neurons (Fig. 9).
Each calculates the weighted sum of its inputs, adds a bias
(b) to it, and then passes the output (n) through a transfer
function, also called activation function.

It is important to note that the weights are initiated with random values at the beginning of training. The model's outputs
are then calculated and compared with the expected results.
The difference between these two values is called loss. The
error then gets backpropagated through the network, and the
weights are updated [11].

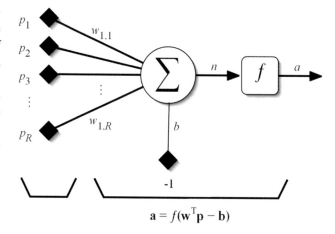

$$\mathbf{a} = f(\mathbf{w}^T\mathbf{p} - \mathbf{b})$$

Fig. 9 Structure of a neuron [34]

The depth (number of hidden layers) and width (number of
neurons in each layer) were determined empirically. The first
hidden layer is a bidimensional convolutional layer with a (3,
3) kernel, this layer is followed by a MaxPooling layer with
a (2, 2) window, and a 0.3 dropout to avoid overfitting. After
that comes a set of two fully connected hidden layers with a
width of 10 and 5, respectively. The transfer function chosen
for each of the hidden layers was rectified linear [34]. It is a
standard activation function that replaces negative values with
zero. For the output layer, only one neuron was necessary, for
the two cases: susceptible or not susceptible to landslides.
The activation function used is called sigmoid. This function
allows us to interpret the output as a probability, this is done by

Table 1 Landslide percentage by lithology class

Lithology class	Landslide percentage
Marl	63.79
Marl and limestone	13.34
Marl, sandstone, and conglomerate	03.68
Sandstone	03.67
Limestone, sandstone, and conglomerate	03.62
Silt	02.73
Limestone	02.08
Marl and sandstone	01.70
Sandstone and silt	01.22
Clay	01.86
Schist and sandstone	00.60
Silt and limestone	00.85
Marl, limestone, and sandstone	00.36
Gypsum	00.11
Flysch	00.39

rescaling the output into a continuous range of values between 0 and 1.

The rest of the hyperparameters were straightforward to setup: binary crossentropy was used to calculate the loss and the weights were updated using the Adaptive Moment Estimation (Adam) optimizer.

3 Results and Discussion

3.1 Model Parameters

Lithology

The vast majority of landslides (64%) have been reported in areas that are dominated by the marl lithology class (Table 1). This sedimentary rock is very frequent in the study area. This type of rock is easily influenced by the various weathering effects that degrade the slopes, which makes marl slopes highly susceptible to landslides.

The marl and limestone class is also affected by a significant number of landslides (12%) although much less important than that of the marl class.

Slope

The slope of the study area varies between a minimum of $0.8°$ in the plains made of silt and a maximum of $55°$ in the steep reliefs of the Rif (Fig. 10).

In rare cases (less than 3%), landslides have been reported in areas with a slope as low as $7°$. The landslide frequency increases gradually as the slopes become steeper, reaching a maximum frequency (32%) around $20°$. After that, landslide frequency takes a downward trend, almost reaching 0% at a slope of $45°$.

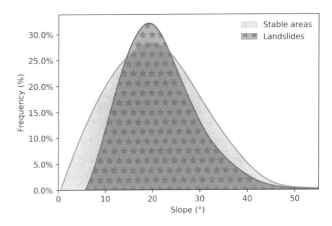

Fig. 10 Slope frequency distribution of the study area

The observed landslide frequency is positively correlated with the overall distribution of the slopes of the study area: the most frequent slope class is also the slope where landslides are most frequent.

Drainage density

The hydrographic network of the study area has well-developed ramifications reaching a density of $5 \ \mathrm{m\,km^{-2}}$ in areas where the water streams are dense (Fig. 11).

A large percentage of landslides have taken place in areas where the density is less than $0.5 \ \mathrm{m\,km^{-2}}$. The rest of the landslides follow a negative density gradient. Areas where the drainage density exceeds $3 \ \mathrm{m\,km^{-2}}$ experience very few landslides.

The hydrographic network of the study area has very few ramifications in the high altitudes. The ramifications become denser as the altitude gets lower. This explains the low frequency of landslides in areas of high drainage density and

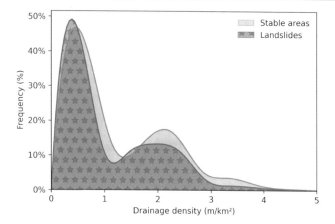

Fig. 11 Drainage density frequency distribution of the study area

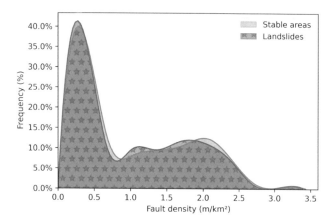

Fig. 12 Fault density frequency distribution of the study area

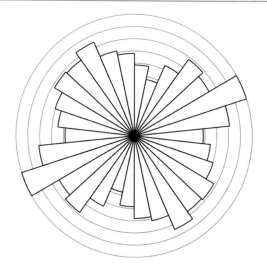

Fig. 13 Dominant fault directions in the study area

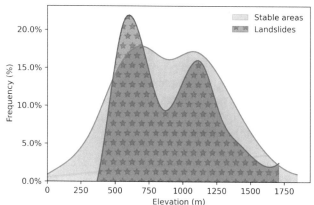

Fig. 14 Elevation frequency distribution of the study area

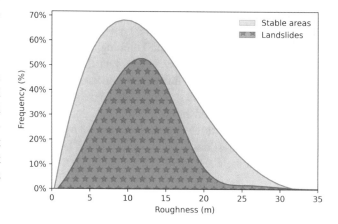

Fig. 15 Roughness frequency distribution of the study area

vice versa. This does not mean that the hydrographic network has no role in the destabilization of the slopes, on the contrary, it becomes a major factor during torrential rains.

Fault density

The analysis of the digitized faults of the study area has shown that 60% of the geological formations have been deformed by a fairly dense network of faults (density between 1 and 2.5 m km^{-2}) (Fig. 12). The dominant fault direction is East-Northeast (Fig. 13). Only 40% of the study area has a fault density $\approx$0.3 m km^{-2}, whereas other areas are with a fault density higher than 3 m km^{-2}. We can also see that landslides are present even in areas not affected by faults.

Hypsometry

The elevation of the study area varies between a minimum of 0 m and a maximum of 1845 m (Fig. 14). High grounds are almost twice as frequent as the plains of the study area.

The elevation frequency distribution for landslides is bimodal: the first peak is located at an altitude of 600 m and the second one is centered around 1100 m.

Roughness

The roughness frequency distribution of landslides as well as of the stable areas is very similar (Fig. 15). The distribution is unimodal, quasi-symmetrical centered around a topographic

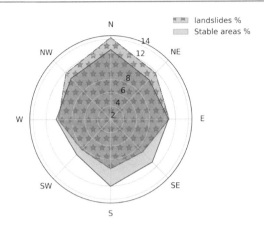

Fig. 16 Aspect frequency distribution of the study area

irregularity of 10–12 m, with landslides appearing at a roughness of 1 m.

Aspect

After plotting the aspect distribution of the study area, we can see that the stable areas have a somewhat balanced aspect distribution, while the majority of landslides that have been reported in the study area favor slopes with a northern exposition (Fig. 16).

North-facing slopes are exposed to the Mediterranean Sea atmosphere and rainy winds. These disturbances play a major role in the destabilization of slopes which could eventually lead to landslides. On the other side, slopes that are oriented toward the south are sheltered by the high reliefs that play the role of natural shields.

3.2   Susceptibility Index

The landslide susceptibility model was trained with a batch size of 16, and the training was stopped at 100 epochs (Fig. 17), after which the validation accuracy did not improve.

The model was evaluated against the out-of-sample data (10% of the data). We found that the accuracy of susceptible areas (92%) was significantly higher than that of non-susceptible (safe) areas (69%). This contrast is mainly due to how we defined safe areas. These areas have been defined as surfaces where no landslides have been mapped on the geological maps. Some safe areas that were chosen in this study could quite possibly be very unstable, and, with a powerful enough trigger factor, could produce landslides. The developed model is therefore less likely to predict false negatives than false positives.

The ANN model was used to generate the susceptibility index in the Aknoul Region (Fig. 18). The output is a probabilistic map with values going from 0 to 1. Using the frequency ratio between susceptibility index and landslides, we found that areas with very high susceptibility index have the most landslides, while areas with very low susceptibility index are affected the least (Fig. 19).

3.3   Discussion

In order to evaluate how our model fairs against different LSM state-of-the-art models, we compared their overall accuracy (Table 2). We found that our model is at the forefront of the state-of-the-art LSM models.

Fig. 17 Evolution of the out-of-sample accuracy for the landslide susceptibility model

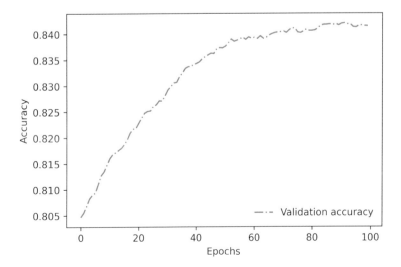

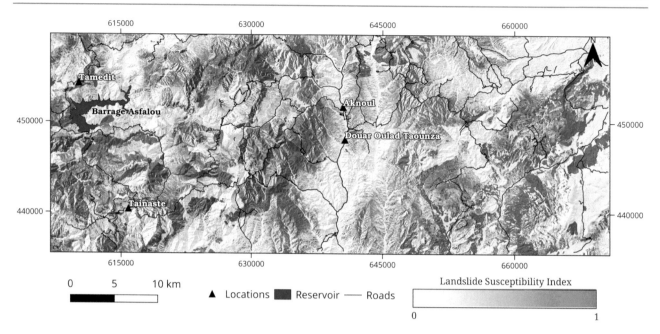

Fig. 18 Landslide susceptibility map of the Aknoul Region

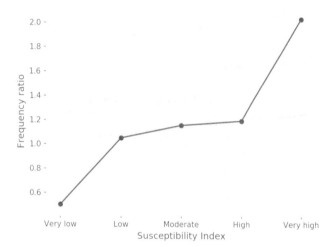

Fig. 19 Frequency ratio by landslide susceptibility index class

We are aware that it is not currently possible to have a fair comparison between the different LSM models, especially since they were developed for specific case studies and use different input data for the training.

A standard annotated landslide dataset is needed to set benchmarks for the LSM models and would allow for an objective comparison between them.

4 Conclusion

In this study, we showcased an efficient landslide susceptibility mapping approach that is based on data derived from geological maps and digital elevation models. After identifying the optimal model architecture to use, we were able to achieve a high accuracy for predicting susceptible areas (92%). The model, however, struggled in properly predicting safe areas (69% accuracy). This is similar to an issue first noted in [11], where the accuracy for the safe areas was significantly lower (more than 40%) than that of the susceptible areas. More researches are needed to be done in order to understand and eventually overcome this issue.

Table 2 Accuracy comparison of multiple LSM models

Model	Accuracy	References
SVM	0.71	[35]
LeNet-5	0.73	[35]
LR	0.75–0.78	[36]
CNN-2D	0.78	[35]
ANN	0.76–0.81	[11]
Our model	0.81	

A landslide susceptibility index map was generated using the ANN model. It showed that landslide occurrence is positively correlated with the susceptibility index.

References

[1] Direction de la Surveillance et de la Prévention des Risques, *Etude pour la réalisation d'une cartographie et d'un système d'information géographique sur les risques majeurs au Maroc* (Etude et mesures les 5 domaines, Rabat, Morocco, 2008)

[2] J.C. Fageollet, *Les mouvements de terrain et leurs préventions* (Collection géographie, 1989)

[3] D. Kempthorne, D.M. Myers, *The Landslide Handbook—A Guide to Understanding Landslides* (U.S. Geological Survey, Reston, Virginia, 2008)

[4] H. Harmouzi, A. Dekayir, M. Rouai, M. Afechkar, Geo-Eco-Trop **42**, 19 (2018), http://www.geoecotrop.be/uploads/publications/pub_421_02.pdf

[5] A. Tribak, Méditerranée **86**, 61 (1997). https://doi.org/10.3406/medit.1997.2991

[6] M. Yazidi, N. Eloutassi, K. Nabih, A. Hammi, A. Yazidi, F. Benziane, Eur. Sci. J. **13**, 46 (2017). https://doi.org/10.19044/esj.2017.v13n12p46

[7] H.B. Wang, K. Sassa, Earth Surf. Process. Landf. **31**, 235 (2006). https://doi.org/10.1002/esp.1236

[8] S. Lee, D.G. Evangelista, Nat. Hazard. Earth Syst. Sci. **06**, 687 (2006). https://doi.org/10.5194/nhess-6-687-2006, www.nat-hazards-earth-syst-sci.net/6/687/2006/

[9] M. Alami Machichi, A. Saadane, L.P. Guth, in *Proceeding 2020 IEEE International Conference of Moroccan Geomatics (Morgeo)* (IEEE, Casablanca, Morocco, 2020), pp. 1–6. https://doi.org/10.1109/Morgeo49228.2020.9121904

[10] T. Ishii, R. Nakamura, H. Nakada, Y. Mochizuki, H. Ishikawa, in *Proceedings of the 14th IAPR International Conference on Machine Vision Applications (MVA)* (2015), pp. 341–344

[11] D. Kawabata, J. Bandibas, Geomorphology **113**, 97 (2009). https://doi.org/10.1016/j.geomorph.2009.06.006

[12] M. Saadi, M. Bensaid, M. Dahmani, G. Choubert, A. Faure-Muret, F. Mégard, T. Mourier, E. Hilali, J. Andrieux, F. Bosson, D. Frizon de Lamotte, G. Roche, Carte géologique du rif al hoceima (1984)

[13] D. Alaoui Mdaghri, M. Bensaid, M. Dahmani, A. Faure-Muret, G. Choubert, J. Braud, J. Morel, D. Frizon de Lamotte, D. Darraz, Carte géologique du rif midar (1994)

[14] M. Fetah, M. Bensaid, M. Dahmani, G. Choubert, A. Faure-Muret, D. Frizon de Lamotte, T. Mourier, F. Besson, M. Blumenthal, J. Andrieux, F. Megard, A. Blondeau, J. Butterlin, H. Feinberg, J. Gay, C. Lorenz, M. Lys, R. Mouterde, J. Sigal, J. Vila, G. Muller, G. Roche, Carte géologique du rif rouadi (1987)

[15] M. Saadi, M. Bensaid, M. Dahmani, G. Choubert, A. Faure-Muret, El Arbi Hilali, J. Houzay, F. de Lamotte, J. Marçais, P. Fallot, G. Suter, J. de Lizaur, A. Marin, A. Jeannette, P. Russo, A. Del Valle, E. Braud-Caire, G. Lecointre, G. Glaçon, R. Busnardo, L. Hottinger, G. Roche, Carte géologique du rif boudinar (1984)

[16] A. Guerraoui, M. Bensaid, M. Dahmani, D. Leblanc, D. Feinberg, H. Lorenz, R. Wernli, M. Septfontaine, Carte géologique du rif ain zohra (1996)

[17] A. Ghissassi, T. Saadi, M. Skalli, A. Boudda, D. Leblanc, H. Feinberg, H. Lorenz, Carte géologique du rif aknoul (1986)

[18] M. Saadi, E. Hilali, M. Bensaid, D. Leblanc, H. Feinberg, H. Lorenz, R. Wernli, D. Leblanc, Carte géologique du rif taineste (1983)

[19] A. Ghissassi, T. Saadi, M. Skalli, E. Hilali, A. Boudda, P. Lespinasse, J. Magné, Carte géologique du rif beni ahmed (1970)

[20] M. Saadi, E. Hilali, A. Boudda, D. Leblanc, H. Feinberg, H. Lorenz, Carte géologique du rif bab el mrouj-taza nord (1978)

[21] A. Ghissassi, T. Saadi, M. Skalli, A. Boudda, D. Leblanc, H. Feinberg, H. Lorenz, Carte géologique du rif ain bou kellal-msoun

[22] P. Rizzoli, M. Martone, C. Gonzalez, C. Wecklich, D. Borla Tridon, B. Brautigam, M. Bachmann, D. Schulze, T. Fritz, M. Huber, B. Wessel, G. Krieger, M. Zink, A. Moreira, ISPRS J. Photogramm. Remote Sens. **132**, 119 (2017). https://doi.org/10.1016/j.isprsjprs.2017.08.008

[23] QGIS, Welcome to the QGIS project! (2020), https://www.qgis.org/en/site/

[24] F. Chollet, et al., Keras (2015), https://keras.io

[25] F. Guzzetti, A. Cesare Mondini, M. Cardinali, F. Fiorucci, M. Santagelo, K.T. Chang, Earth Sci. Rev. **112**, 42 (2012)

[26] M. Kirat, Revue Interdisciplinaire (2017)

[27] L. Ait Brahim, M. Bousta, I.A. Jemmah, I. El Hamdouni, A. ElMahsani, A. Abdelouafi, F. Sossey alaoui, I. Lallout, in *MATEC Web of Conferences* (EDP Sciences, 2018). https://doi.org/10.1051/matecconf/201814902084

[28] D.M. Mark, Geografiska Annaler 165–177 (1975)

[29] J. McKean, J. Roering, Geomorphology 331–351 (2004)

[30] M. Van Den Eeckhaut, J. Poesen, G. Verstraeten, V. Vanacker, J. Moeyersons, J. Nyssen, L.P.H. van Beek, Geomorphology 351–363 (2005)

[31] A.M. Booth, J.J. Roering, J.T. Perron, Geomorphology 132–147 (2009)

[32] Y. LeCun, Y. Bengio, G. Hinton, Nature **521**, 436 (2015). https://doi.org/10.1038/nature14539

[33] F. Chollet, *Deep Learning with Python* (Manning Publications Co., Shelter Island, NY, 2018)

[34] M. Parizeau, *Réseaux de neurones* (Université LAVAL, 2006)

[35] Y. Wang, Z. Fang, H. Hong, Sci. Total Env. **666**, 975 (2019). https://doi.org/10.1016/j.scitotenv.2019.02.263

[36] A.N. El Fahchouch, L. Ait Brahim, O. Raji, in *MATEC Web Conference* (2018). https://doi.org/10.1051/matecconf/201814902055

Lithological Mapping for a Semi-arid Area Using GEOBIA and PBIA Machine Learning Approaches with Sentinel-2 Imagery: Case Study of Skhour Rehamna, Morocco

Imane Serbouti, Mohammed Raji, and Mustapha Hakdaoui

Abstract

Accurate and reliable lithological mapping through satellite-borne remote sensing data and image classification approaches has a critical role since it can automatically and promptly identify lithological units over large areas. Most available Pixel-Object Based comparative classification studies have been applied to land use land cover (LULC) studies; however, this research aims to evaluate and compare the performance of these digital classification methods in the field of geological mapping in semi-arid areas, by integrating spectral bands and neo-bands, particularly the Minimum noise fraction (MNF) and the principal component analysis (PCA), of Sentinel-2A satellite imagery, to map the southern of Skhour Rehamna which is located at the western Moroccan Meseta. The analysis results from two different methods, namely, pixel-based image analysis (PBIA) with k-nearest neighbour (K-NN) and Random Forest (RF) machine learning algorithms (MLAs), and Geographic Object-Based Image Analysis (GEOBIA) were assessed and compared. PBIA method involved selection of training areas whether it was k-NN or RF MLAs, and produced lithological maps that exhibit "salt and pepper" effects as well as problems associated to delineating accurate lithological boundaries, while GEOBIA approach involved multi-resolution segmentation step where scale, shape and compactness parameters should be adjusted as accurate as possible, in order to segment the image into homogeneous and meaningful regions so that the resulted samples were classified using Standard Nearest Neighbour algorithm. Therefore, the resulting lithological maps were assessed by comparing both techniques using confusion matrix, overall accuracy (OA) and Kappa coefficient (K). The results show that the GEOBIA approach had higher overall agreement (83.46% OA and 0.76 K) than RF (81.92% OA and 0.72 K) and k-NN (80.79% OA and 0.70 K) PBIA approaches. Overall, the results clearly indicate the potential of GEOBIA technique for lithological mapping applications to produce more realistic maps.

Keywords

Lithological mapping • Sentinel-2 imagery • Skhour Rehamna • MLAs • RF • k-NN • PBIA • GEOBIA

1 Introduction

Nowadays, the use of remotely sensed spectral data has become a very powerful and popular technique for geological mapping, specifically in arid and semi-arid areas [1–4], due to its advantages in terms of cost efficiency, accuracy and time consuming in discriminating lithological units automatically over vast regions.

In the data side, the development of multi-spectral remote sensing technology has revolutionized the techniques to extract information about earth's surface [5]. Optical remote sensing imagery, including both spaceborne and airborne sensors, differs in spectral, spatial and temporal resolutions. Since the selection of suitable earth observation (EO) datasets is considered as the first essential step for a successful image classification [6–8], the Sentinel-2 multispectral imager (MSI) developed by the European Space Agency (ESA) have shown a great potential for lithological mapping and mineral exploration in last decades, due to its high spatial and spectral resolutions compared to Landsat and SPOT sensors datasets especially in the VNIR region [9, 10].

Previous studies evaluate the impact of remotely sensed data and the fusion of SAR and optical datasets on lithological mapping [11], otherwise the selection of a suitable digital image classification is also a fundamental process to

I. Serbouti (✉) · M. Raji · M. Hakdaoui
Department of Geology, Laboratory of Applied Geology,
Geomatic and Environment, Faculty of Sciences Ben M'Sik,
Hassan II University of Casablanca, Casablanca, Morocco

© The Author(s), under exclusive license to Springer Nature Switzerland AG 2022
F. Barramou et al. (eds.), *Geospatial Intelligence*, Advances in Science, Technology & Innovation,
https://doi.org/10.1007/978-3-030-80458-9_11

produce and update geological maps by relating pixel values to lithological units present on earth surface. Most of the approaches used to produce lithological map of a region use either pixel-based image analysis (PBIA) [12, 13] or Geographic Object-Based image analysis (GEOBIA), also termed as Object-Based image analysis (OBIA) [14].

The most common approach utilized for this target is PBIA approach, and it consists on analysing and distinguishing the closest match between spectral information of each pixel and the single ground class apart from [15] without examining the contextual, textural and spatial properties associated to the pixel of interest [16]. A wide range of classification methods has been applied for this purpose; it can be categorized by its statistical underlying assumptions (e.g., parametric vs. non-parametric), the way in which elements are classified (i.e., per-pixel, and subpixel), or the requirement of collecting representative endmember samples (e.g., supervised vs. unsupervised) [17]. The use of machine learning algorithms (MLAs) is influenced by many factors, including the selection of the right training samples, the choice of the ideal features and optimization of training parameters [18], which leads per-pixels classification algorithms more challenging [19, 20].

Numerous geoscientific studies have used PBIA MLAs especially in lithological mapping and mineral exploration [17, 21].

In contrast to PBIA classification methods that assign a class directly to individual pixels and cause problems associated with heterogeneity of earth surface and solar illumination angle that occur some drawbacks such as "salt and pepper" noise and topographic unfavourable effects [22] especially with high resolution images such as Sentinel-2 image with 10 m spatial resolution, GEOBIA approach has been developed to improve the deficiency of PBIA by introducing in addition to spectral characteristics, the spatial textural properties such as texture, shape, colour, size and association between the neighbouring objects [23], and these by using an additional critical stage in the classification process of this approach which is the multiresolution segmentation technique that aggregate like-pixels into homogeneous meaningful objects with similar spectral, textural and spatial information, and then assign the category of each feature by using classifiers, in this study the standard nearest neighbour classifier (SNN) was applied.

Several researches have applied GEOBIA approach for a variety of applications, including land use land cover mapping (LULC) [24, 25], lithological mapping [26], change detection [27], landform mapping [28], urban mapping [29], crop and vegetation classification [30, 31], with many studies demonstrated that GEOBIA approach produced a higher thematic classification accuracies than the traditional PBIA approaches [32, 33].

This study has been structured into two parts, the first consist on evaluating the performance of the supervised non-parametric PBIA machine learning algorithms, including Random Forest (RF) and k-Nearest Neighbour (k-NN), while the second part provides a more complete evaluation of GEOBIA and PBIA classification approaches for lithological mapping in the southern part of Skhour Rehamna, situated in the western Moroccan Meseta, using Sentinel-2 imagery.

2 Location and Geological Settings of the Study Area

Skhour Rehamna is an inlier of the Paleozoic and Paleoproterozoic basement that forms the Hercynian Rehamna massif (Central Morocco) to the north and the Jebilet to the south. In the division of the Hercynian chain of Morocco, this region belongs to the western Moroccan Meseta, where erosion dissects the sub-tabular Cretaceous-Eocene cover of the Gantour Plateau, more precisely located on approximately 100 km from Marrakech, crossed from north to south by the A7 highway and the No. 9 principle road linking Casablanca to Marrakech [34].

The focus of this research is a region along the southern of the Paleozoic massif of Skhour Rehamna that lies between the meridians 7°54′55″ and 7°43′50″ west and the parallels 32°22′30″ and 32°14′39″ north, as highlighted in Fig. 1 below, in order to analyse more precisely the results obtained.

The study area (Fig. 2) is made up of stacked mica schist formations attributed to the Devonian (the Unit of Ouled Hassine) [35] that correspond to a pelitic series with six intercalations of quartzites and Metabasite and to the Paleoproterozoic (Lalla Tittaf Formation) [36], which contain metapelites and semipelites with intercalations of metabasites, orthogneiss, calcschists and marbles between the two lies the unit of Dalaat el Kahlat, which the age remains unknown [34]. The small granitic intrusions of Ras el Abiod are arenized from Pliovi lafranchien and expressed at the surface as a large area of thermal metamorphism. The Maastrichtian is directly transgressive on the mica schists and the Permian in the southern part of the map region, creating a cuesta clearly dominating the Paleozoic inlier. This is the plateau where the phosphates of Benguerir are mined [34].

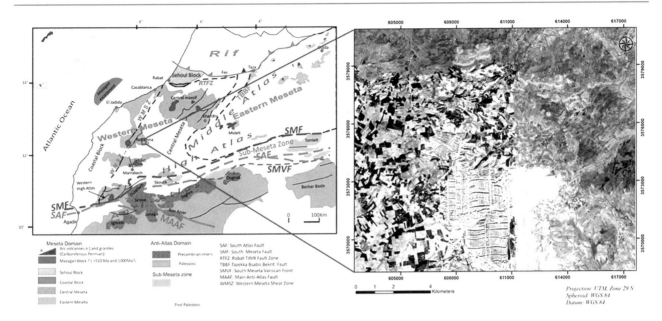

Fig. 1 Location of the southern of Skhour Rehamna (Google Earth, resolution 0.5 m) on the map of the geological domains of Morocco (modified by Michard et al. 2010)

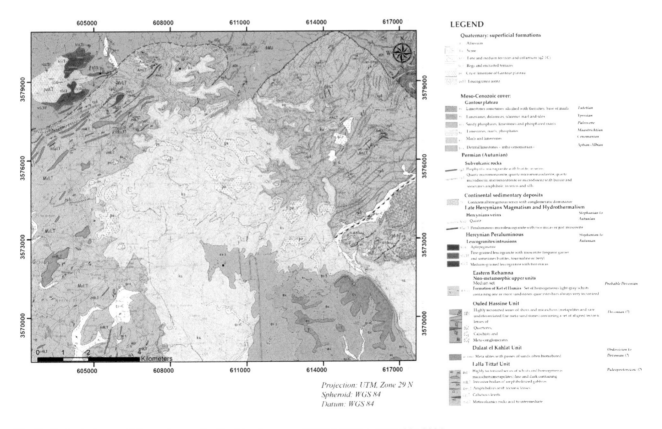

Fig. 2 Geological map of the study area (realized by the group BRGM-CID) published in 2004

3 Materials and Methods

3.1 EO Datasets Properties and Pre-processing

The satellite imagery source used in this study is Sentinel-2A product carry on board multispectral imaging instruments (MSI) with 13 wide-swaths spectral bands in the visible near infrared (VNIR) and short-wave infrared (SWIR) [37] and high to moderate spatial resolution ranging from 10 to 60 m [38, 39]. The VNIR spectral bands have a spatial resolution of 10 m which makes this product involve the potential for detailed exploration of earth surface, the infra-red bands have 20 m, and the three atmospheric corrections have 60 m spatial resolution [40].

In the following study we opted Sentinel-2A (Level 1C) imagery acquired on 29 October 2017. In order to achieve the level desired by the user, Sentinel-2 MSI products undergo multiple stages of processing; for this purpose the ESA Sen2Cor plugin available on the Sentinel Application Platform (SNAP) [37] was used to process reflectance image bands from Level 1C Top of Atmosphere (TOA) product, to Level 2A Bottom of Atmosphere (BOA) Sentinel-2 imagery, by applying Terrain and atmospheric corrections. Due to the low spatial resolution (60 m) and the sensitivity to the clouds and aerosol, spectral bands 1, 9 and 10 were omitted in this research. The remained bands with spatial resolution of 20 m (5, 6, 7, 8a, 11, 12) were cubically resampled to $10*10m^2$ spatial resolution to reach the same resolution as VNIR bands (2, 3, 4 and 8). Finally, all the bands were re-projected to the UTM (Universal Transverse Mercator projection) WGS84 in zone 29 N coordinate system.

3.2 Methodology

At a time when many innovative classification approaches were already produced, the Sentinel-2 satellite was launched. These approaches are based on pixels [41, 42] and objects [14, 43, 44]. To find the optimal method for the classification assessment of lithological units in the selected region using Sentinel-2 imagery, two typical machine learning algorithms, particularly RF and k-NN, were commonly applied and compared to GEOBIA approach. For the purpose of ensuring more diagnostic spectral features of the exposed rock units, numerous neo-bands extracted from Eigen-space-based algorithms in particular, the Minimum noise fraction (MNF) and the principal component analysis (PCA) were layer-stacked to Sentinel-2 spectral bands.

An outline of the methodology used in this study is demonstrated in the flow diagram (Fig. 3). However, the following sections described the data processing details, classification techniques applied in this study and subsequent statistical evaluations.

Spectral Features Analysis. In general, multispectral limited channels provide a collection of mixed-pixels representing undistinguishable ground features [45–47]. Therefore, this challenge is overcome through dimensionality reduction of MSI bands using principal component analysis (PCA) and Minimum noise fraction (MNF) [48].

Principal Component Analysis (PCA). This transformation is a multivariate statistical and data reduction procedure, commonly employed for geological mapping [49–53]. In order to highlight and enhance spectral information related to specific rock unit [54], PCA can be applied to MSI

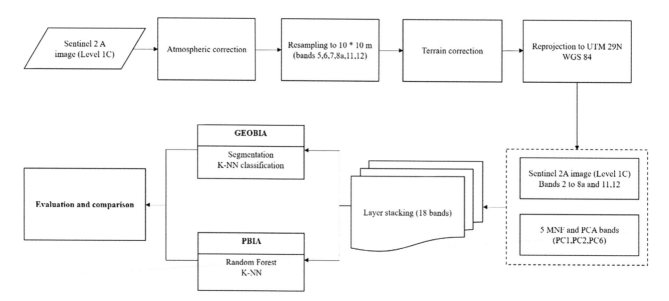

Fig. 3 Workflow of the methodology applied in this study

Fig. 4 Colour composite of the PCs 1, 2, and 6 of the Sentinel2 imagery

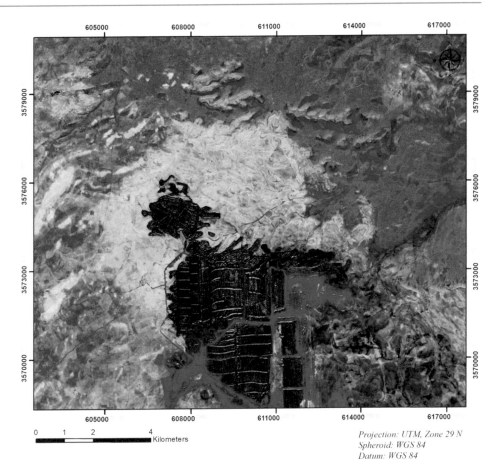

```
0    1    2         4
                  Kilometers
```

Projection: UTM, Zone 29 N
Spheroid: WGS 84
Datum: WGS 84

datasets by transforming the original and high dimensional set of features to an uncorrelated lower dimension output bands through the calculation of covariance matrix, eigenvector and eigenvalue pairs as well as data orthogonal projection [55]. The dimensionality of the datasets is reduced by eliminating redundant data by extracting maximum information, and the first and the second PCs include the majority percentage of the scene variance in the data and succeeding component bands with a decreasing percentage of the variance [22]. Hence, we have selected three PCs in this research (PC1, PC2 and PC6) to generate a colour composite map (Fig. 4) that enable to better discriminate the lithological units and trace training polygons for the classification approaches used in this study.

Minimum Noise Fraction (MNF). In order to reduce the residual noise of reflectance images and showcase homogeneous surfaces, Minimum Noise Fraction (MNF) technique [56, 57] was carried out. It is a wildly known Eigenvector procedure for multispectral and hyperspectral image, based on covariance structure of imagery noise. This algorithm consists of two successive PCA rotation which transforms data containing spectral distortion into new components sorted by image quality, with regularly increasing noise levels. The first one accounts for the

covariance matrix to estimate the noise in the data in order to decorrelate and resize the noise, and the second rotation is based on a standard PCA transform to create several components that contain noise-whitened data. This results in denoising and identifying the components to keep those with useful information [58].

A visual study of the first three components from the MNF (containing more than 99% of the total information) allows discrimination between different surfaces in the study area (Fig. 5).

PBIA Lithological Mapping. One of the most traditional classification methods used for Sentinel-2 imagery is the pixel-based MLAs, which allocate any pixel to a specific category, taking into account the spectral characteristics of the training samples that group a set of pixels representing the same class [59], the thing that makes selecting suitable training sample as the one of the most crucial step of PBIA classification approaches. The literature shows that among the MLAs used for classifying the lithological units using multispectral datasets, RF and k-NN are the most common MLAs applied for this purpose [60, 61].

Random Forest (RF). The first MLA implemented in this study is the Random Forest classifier (RF) developed by Breiman [62] and applied for remote sensing image

Fig. 5 Colour composite of the
MNF bands 1, 2 and 3 of the
Sentinel2 image

Projection: UTM, Zone 29 N
Spheroid: WGS 84
Datum: WGS 84

classification by Pal [63]. It is a supervised non-parametric classification algorithm, which provides a group of tree classifiers that choose the majority vote class to assign a label for each pixel to be classified based on the partition of the results from multiple decision trees (DT). Randomness is introduced by randomly requiring a predefined number of characteristic parameters (mtry) and the input variables for each decision tree (ntree), by setting the input variables for splitting at each node in the DT and bagging; the latter technique, also known as bootstrap aggregation, is used to select training samples available for every tree [62]. Thus, each tree in the forest votes for the final classes produced by the forest.

RF performs greater than the other MLAs using numerous techniques such as bagging and boosting [64]. As RF is sensitive to the training samples, their spatial dispersion must be increased to improve classification results.

Furthermore, several studies have even proven to achieve optimal accuracies and vital lithological maps using RF algorithm compared to other MLAs like, Naive Bayes, k-Nearest Neighbours and Artificial Neural Networks [60], support vector machine [65, 66].

k-Nearest Neighbour (k-NN). The second algorithm applied in this research is k-NN classifier, and it is one of the most simple, popular, and instance-based non-parametric machine learning algorithms [67]. During classification, individual test instance that is nearest to k neighbouring training sets is in a feature space, based on a Euclidian distance metric function:

$$dE(x, y) = \sum_{i=1}^{N} \sqrt{x_i^2 + y_i^2} \qquad (1)$$

where x and y are histograms in X = Rm (and m is dimensionality of the image). Figure 6 shows the process of KNN classification [68]. Predictions are assigned by the majority vote among its k nearest neighbour samples [69–71]. If $k = 1$, then the object is simply assigned to the same class of the object nearest to it. k must be generally an odd integer if the number of classes is two. As a very low value of K lead to noisy and cause effects of misfits in the model, as well as high k can lead to smoother decision boundaries and instability in the model, appropriate values must be selected by trial and error [72].

GEOBIA Lithological Mapping. In contrast to the PBIA approach, GEOBIA [73] is based on information extracted from a group of similar pixels, according to their spatial spectral and textural information, that is called image objects, which plays an important role in the classification by

Fig. 6 The MRS result on the background colour composite image of the first three MNF bands

taking into consideration spectral content, size as well as the shape [74]. In this approach the image must be segmented into homogeneous and meaningful objects (step1) before the classification process (step2).

Multi-Resolution Segmentation (MRS). The Multiresolution segmentation (MRS) algorithm is recognized as the first and the most crucial step in GEOBIA approach because its outcomes influence directly all the following process [75]. MRS successively implements a bottom-up region merging technique that begins at random points with single pixels objects and then merges them into larger and real-world segments depending on the homogeneity criterion [76]. The purpose of this stage is to create real-world objects that would be classified according to their contextual, textural, spatial as well as spectral homogeneity. The MRS method's outcome is based on four parameters, namely, layer weight, compactness, shape and scale parameter (SP). Compactness is known as the weight of smoothness criterion, likewise the shape-colour criteria refers to spectral information of an object, whereas SP defines the maximum

heterogeneity of the image objects [77]. The result of the MRS is illustrated in Fig. 6.

Classification Algorithm. The second and last stage in GEOBIA approach is selecting a set of feature vector to differentiate between the target classes and create connectivity between real-world classes and the image objects to apply a suitable classification rule. In this study, the classification of image objects was carried out by standard Nearest Neighbour (NN) classifier. It consists of searching for the appropriate training sample in the feature space for each object [76].

Accuracy assessment. In order to evaluate the classification accuracy of for all the classification methods in this research, the resultant lithological maps were assessed by comparing them the digitalized geological map of the study region using the confusion matrix [78]. Several measurements, including overall accuracy (OA), commission and omission errors, and a kappa coefficient (K), were calculated to identify the potential of each classification approach.

(a)   **(b)**

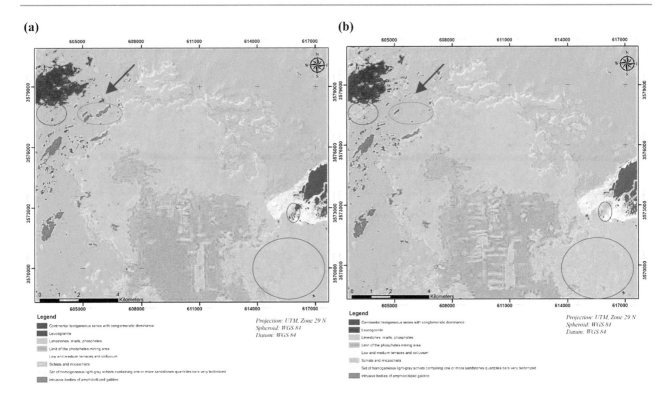

Fig. 7 The resultant lithological maps using PBIA MLAs: **a** RF; **b** k-NN

3.3 Results and Discussion

The lithological map obtained from PBIA, namely RF and K-NNMLAs as well as GEOBIA approach, was illustrated and assessed in the sub-sections below.

PBIA Results. The resultant lithological maps developed using Sentinel-2 imagery for both pixel-based MLAs are described in Fig. 7.

In addition, a general comparison between the two Pixel-based MLAs (Fig. 7) reveals that the k-NN method (Fig. 7b) showed many facies that are poorly classified, for instance, the circles with orange and magenta colours that show the apparition of some misclassified classes that appears in k-NN approach, in addition to the leucogranite, marked by a red arrow, that has been assigned as Limestone, marls, phosphate in k-NN approach, and the continental terrigeneous series with conglomeratic dominance (CC), that are indistinguishable in k-NN approach as demonstrated by the blue circle, finally, as illustrated by the green circle some of the intrusive bodies of amphibolitized gabbro (AG), have been clearly manifested in RF MLA (Fig. 7a).

About the classification accuracy, RF performs much better for lithological mapping (OA = 81.92% and Kappa coefficient = 0.72) compared to k-NN algorithm.

GEOBIA Results. Unlike the PBIA approaches, GEOBIA shows homogeneous classes and reduces all the problems related to the misclassified pixels as well as salt and pepper artifacts since it performs by not only taking into account spectral properties, but also the shape, texture, and geometry of objects during the process of classification. Furthermore, as shown in Fig. 8, the GEOBIA technique has greater potential to generate lithological maps in which the overall accuracy of the classification results (OA = 83.46%, Kappa coefficient = 0.76) outperformed PBIA machine learning algorithms.

Overall Comparison. The confusion matrix has been used in this study to evaluate the efficiency of the classification accuracy for the geological maps obtained using both MLAs of the PBIA approach (RF and k-NN) and GEOBIA technique. Therefore, the known pixels from the digitalized lithological map of the study area were used as reference data. Besides, the digitalized geological map (Fig. 9) of the study area is depicted into eight general classes: Continental terrigeneous series with conglomeratic dominance (CC), Leucogranite (LG), Limestones, marls, phosphates (LMP), Limit of the phosphates mining area (LPMA), Low and medium terraces and colluvium (LC), Schists and micaschists (SM), Set of homogeneous light-gray schists

Fig. 8 The resultant lithological maps using GEOBIA approach

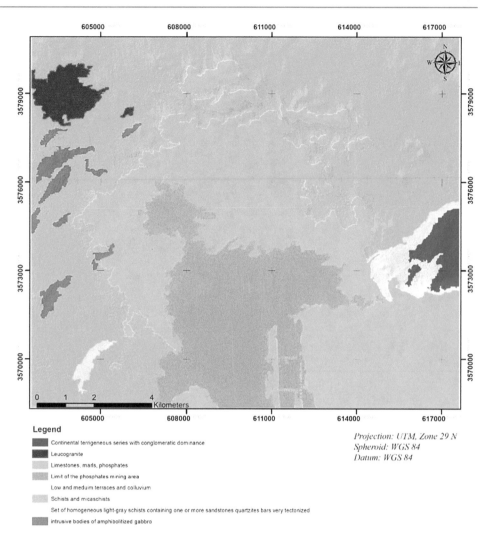

Legend

■ Continental terrigeneous series with conglomeratic dominance
■ Leucogranite
Limestones, marls, phosphates
Limit of the phosphates mining area
Low and medium terraces and colluvium
Schists and micaschists
Set of homogeneous light-gray schists containing one or more sandstones quartzites bars very tectonized
■ Intrusive bodies of amphibolitized gabbro

Projection: UTM, Zone 29 N
Spheroid: WGS 84
Datum: WGS 84

containing one or more sandstones quartzites bars very tectonized (SQ) and intrusive bodies of amphibolitized gabbro (AG).

Tables 1, 2, and 3 display the confusion matrix of PBIA (RF, K-NN) and GEOBIA approaches that is derived by comparing the corresponding classes to the reference samples.

General comparison of all classes shows that many lithological units are misclassified especially for intrusive bodies of amphibolitized gabbro (AG) and Low and medium terraces and colluvium (LC) for both PBIA MLAs (Tables 1 and 2), and these could be demonstrated by omission and commission errors that are greater than those of GEOBIA approach (Table 3). However, the overall accuracy and kappa coefficient for each method are shown in Fig. 10.

The misclassified classes, as well as salt and pepper artifacts caused by the effect of mixing pixel problem in PBIA algorithms, led to the lowest overall and kappa values; however, GEOBIA approach improved the results by achieving the highest accuracy statistics.

4 Conclusions

Finding the optimal classification method is the most critical step for geological mapping; for this purpose, this study is devoted to evaluate different approaches including pixel and object-based image analysis, in order to select the most accurate approach for mapping lithological units in semi-arid areas, where Skhour Rehamna was chosen as a case study.

The lithological mapping was successfully achieved by evaluating the performance of GEOBIA and PBIA approaches using spectral channels and neo-bands of Sentinel-2A imagery. However, the overall statistics of this research obviously indicate that the GEOBIA approach has considerable potential and advantages for generating more realistic and detailed lithological maps also acquiring lithological information and properly classifying all lithological units by reducing all the problems encountered while using PBIA MLAs.

Fig. 9 Digitalized geological map of study area

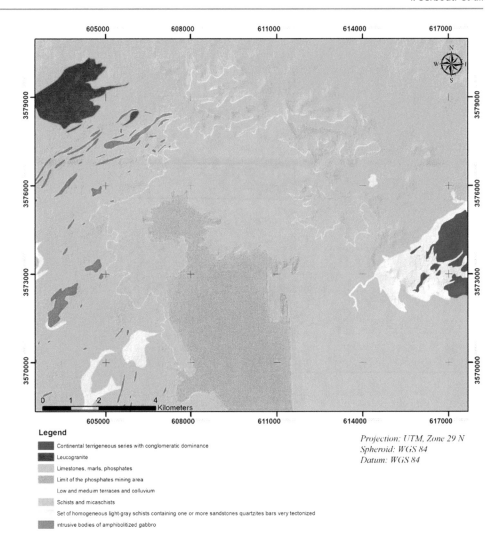

Legend

Continental terrigeneous series with conglomeratic dominance

Leucogranite

Limestones, marls, phosphates

Limit of the phosphates mining area

Low and medium terraces and colluvium

Schists and micaschists

Set of homogeneous light-gray schists containing one or more sandstones quartzites bars very tectonized

Intrusive bodies of amphibolitized gabbro

Projection: UTM, Zone 29 N
Spheroid: WGS 84
Datum: WGS 84

Table 1 Confusion Matrix of PBIA RF Classification

Pixel-based classification *PBIA-RF*

		Reference data									
		LC	*CC*	*AG*	*LG*	*SQ*	*LMP*	*SM*	*LPMA*	Total	Commission error (%)
Classified data	*LC*	6602	213	69	0	646	0	0	3376	10,906	39.46
	CC	1487	15,587	75	0	3339	8	0	2998	23,494	33.66
	AG	158	0	6313	318	0	1360	0	15,488	23,637	73.29
	LG	0	63	0	28,377	128	338	0	7109	36,015	21.21
	SQ	1595	3279	0	0	17,967	81	50	144	23,116	22.27
	LMP	4262	691	395	705	2689	471,323	20,470	63,262	563,797	16.40
	SM	977	0	0	383	25	134,885	209,522	7050	352,842	40.62
	LPMA	25,683	3012	17,342	7718	4534	11,117	549	822,372	892,327	7.84
	Total	40,764	22,845	24,194	37,501	29,328	619,112	230,591	921,799	1,926,134	
	Omission error (%)	83.80	31.77	73.91	24.33	38.74	23.87	9.14	10.79		

Table 2 Confusion matrix of PBIA K-NN classification

Pixel-based classification *PBIA-KNN*

		Reference data									
		LC	CC	AG	LG	SQ	LMP	SM	LPMA	Total	Commission error (%)
Classified data	LC	5835	213	0	0	61	0	0	608	6717	13.13
	CC	1492	15,276	75	0	2926	8	0	3227	23,004	33.59
	AG	158	0	5362	318	0	1360	0	14,286	21,484	75.04
	LG	0	62	0	27,237	122	338	0	5444	33,203	17.97
	SQ	1595	2526	0	0	17,498	81	63	144	21,907	20.13
	LMP	5070	691	464	1683	2756	464,421	24,674	78,063	577,822	19.63
	SM	1034	0	0	700	25	141,835	205,301	4773	353,668	41.95
	LPMA	25,580	4077	18,293	7563	5940	11,069	553	815,252	888,327	8.23
	Total	40,764	22,845	24,194	37,501	29,328	619,112	230,591	921,797	1,926,132	
	Omission error (%)	85.69	33.13	77.84	27.37	40.34	24.99	10.97	11.56		

Table 3 Confusion matrix for GEOBIA approach

Object-based classification *GEOBIA*

		Reference data									
		LC	CC	AG	LG	SQ	LMP	SM	LPMA	Total	Commission error (%)
Classified data	LC	580	33	0	0	42	7	360	0	1022	43.25
	CC	70	810	0	0	384	0	0	0	1264	35.92
	AG	1	0	391	2	0	22	717	0	1133	65.49
	LG	0	0	1	1164	0	0	210	0	1375	15.35
	SQ	76	132	0	0	710	19	0	0	937	24.23
	LMP	208	17	11	0	63	20,682	2710	769	24,460	15.45
	SM	914	0	991	390	0	548	30,992	0	33,835	8.40
	LPMA	16	0	8	0	0	3563	338	8365	12,290	31.94
	Total	1865	992	1402	1556	1199	24,841	35,327	9134	76,316	
	Omission error (%)	68.90	18.35	72.11	25.19	40.78	16.74	12.27	8.42		

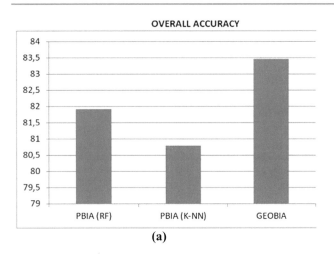
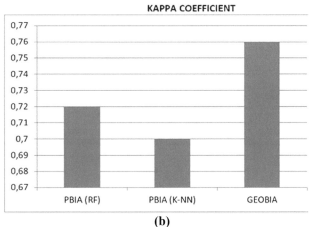

Fig. 10 **a** Overall accuracies (OA) and **b** Kappa coefficient comparison of RF, k-NN PBIA and GEOBIA approaches

References

1. F. Masoumi, T. Eslamkish, A.A. Abkar, M. Honarmand, J.R. Harris, Integration of spectral, thermal, and textural features of ASTER data using Random Forests classification for lithological mapping. J. African Earth Sci. **129**, 445–457 (2017)
2. W. Ge, Q. Cheng, Y. Tang, L. Jing, C. Gao, Lithological classification using Sentinel-2A data in the Shibanjing ophiolite complex in Inner Mongolia, China. Remote Sens. **10**(4), 2018
3. A. Rezaei, H. Hassani, P. Moarefvand, A. Golmohammadi, Lithological mapping in Sangan region in Northeast Iran using ASTER satellite data and image processing methods. Geol. Ecol. Landscapes **4**(1), 59–70 (2020)
4. S. Gad, T. Kusky, Lithological mapping in the Eastern Desert of Egypt, the Barramiya area, using Landsat thematic mapper (TM). J. African Earth Sci. **44**(2), 196–202 (2006)
5. T. Jucker et al., Allometric equations for integrating remote sensing imagery into forest monitoring programmes. Glob. Chang. Biol. **23**(1), 177–190 (2017)
6. S.R. Phinn, C. Menges, G.J.E. Hill, M. Stanford, Optimizing remotely sensed solutions for monitoring, modeling, and managing coastal environments. Remote Sens. Environ. **73**(2), 117–132 (2000)
7. P. Of and I. Creation, Chapter 2 Selection of Remotely Sensed Data.
8. S.R. Phinn, A framework for selecting appropriate remotely sensed data dimensions for environmental monitoring and management. Int. J. Remote Sens. **19**(17), 3457–3463 (1998)
9. Y. Murayama, M. Ranagalage, remote sensing Sentinel-2 data for land cover/use mapping: a review. Remote Sens. **2291**(12), 14 (2020)
10. F.D. van der Meer et al., Multi- and hyperspectral geologic remote sensing: a review. Int. J. Appl. Earth Obs. Geoinf. **14**(1), 112–128 (2012)
11. I. Serbouti, M. Raji, M. Hakdaoui, *Integrating GEOBIA and multisource remote sensing images to lithological mapping: case study of Skhour Rehamna, Morocco* (2020) p. 62
12. M. Kovačević, B. Bajat, B. Trivič, R. Pavlovič, Geological units classification of multispectral images by using support vector machines. Int. Conf. Intell. Netw. Collab. Syst. INCoS **2009**, 267–272 (2009)
13. P.Gong, Integrated analysis of spatial data from multiple sources: using evidential reasoning and artificial neural network techniques for geological mapping.pdf. Photogramm. Eng. Remote Sens. **62** (5), 513–523
14. S. Grebby, E. Field, K. Tansey, Evaluating the use of an object-based approach to lithological mapping in vegetated terrain. Remote Sens. **8**(10), 2016
15. E.O. Makinde, A.T. Salami, J.B. Olaleye, O.C. Okewusi, Object based and pixel based classification using Rapideye satellite imager of ETI-OSA, Lagos, Nigeria. Geoinform. FCE CTU **15**(2), 59–70 (2016)
16. P. Blaschke, T., Lang, S., Lorup, E., Strobl, J. Zeil, Object-oriented image processing in an integrated GIS/remote sensing environment and perspectives for environmental applications. Environ. Inf. Planning, Polit. Public **2**, 555–570 (2000)
17. S. Asadzadeh, C. R. de Souza Filho, A review on spectral processing methods for geological remote sensing. Int. J. Appl. Earth Obs. Geoinf. **47**, 69–90 (2016)
18. D. Michie, Memo functions and machine learning. Nature **218**, 19–22 (1968)
19. C. Cortes, L.D. Jackel, W.-P. Chiang, Limits on Le Back Thesis arming Machine Accuracy Imposed by Data Quality," Neural Inf. Process. Syst., pp. 239–246, 1995.
20. G.E.A.P.A. Batista, R.C. Prati, M.C. Monard, A study of the behavior of several methods for balancing machine learning training data. ACM SIGKDD Explor. Newsl. **6**(1), 20 (2004)
21. M.S. Tehrany, B. Pradhan, M.N. Jebuv, A comparative assessment between object and pixel-based classification approaches for land use/land cover mapping using SPOT 5 imagery. Geocarto Int. **29** (4), 351–369 (2014)
22. T. M. Lillesand, Remote Sens. Image Interpret. **7**(9), (2015)
23. T. Blaschke et al., Geographic object-based image analysis: towards a new paradigm. ISPRS J. Photogramm. Remote Sens. **87**, 180–191 (2014)
24. D.G. Goodin, K.L. Anibas, M. Bezymennyi, Mapping land cover and land use from object-based classification: an example from a complex agricultural landscape. Int. J. Remote Sens. **36**(18), 4702–4723 (2015)
25. O. Akcay, E. O. Avsar, M. Inalpulat, L. Genc, A. Cam, Assessment of segmentation parameters for object-based land cover classification using color-infrared imagery. ISPRS Int. J. Geo-Inform. **7**(11) (2018)
26. S. Imane, R. Mohamed, H. Mustapha, A comparison of GEOBIA Vs PBIA machine learning methods for lithological mapping using Sentinel 2 imagery: case study of Skhour Rehamna, Morocco, in

Proceedings of the 2020 IEEE International Conference on Moroccan Geomatics, MORGEO 2020 (2020), pp. 2–7

27. J. Im, J.R. Jensen, J.A. Tullis, Object-based change detection using correlation image analysis and image segmentation. Int. J. Remote Sens. **29**(2), 399–423 (2008)

28. L. Drăguţ, T. Blaschke, Automated classification of landform elements using object-based image analysis. Geomorphology **81** (3–4), 330–344 (2006)

29. R. Momeni, P. Aplin, D.S. Boyd, Mapping complex urban land cover from spaceborne imagery: the influence of spatial resolution, spectral band set and classification approach. Remote Sens. **8**(2) (2016)

30. V. Lebourgeois, S. Dupuy, É. Vintrou, M. Ameline, S. Butler, A. Bégué, A combined random forest and OBIA classification scheme for mapping smallholder agriculture at different nomenclature levels using multisource data (simulated Sentinel-2 time series, VHRS and DEM). Remote Sens. **9**(3) (2017)

31. M. Belgiu, O. Csillik, Sentinel-2 cropland mapping using pixel-based and object-based time-weighted dynamic time warping analysis. Remote Sens. Environ. **204**, 509–523 (2018)

32. S.W. Myint, P. Gober, A. Brazel, S. Grossman-Clarke, Q. Weng, Per-pixel vs. object-based classification of urban land cover extraction using high spatial resolution imagery. Remote Sens. Environ. **115**(5), 1145–1161 (2011)

33. I. L. Castillejo-González et al., Object- and pixel-based analysis for mapping crops and their agro-environmental associated measures using QuickBird imagery. Comput. Electron. Agric. **68** (2), 207–215 (2009)

34. Y.N. Baudin, P. Chevremont, P. Razin, T.M. And, D. Andries, C. Hoepffner, D. Thieblemont, E.-M. Chihani, *Carte Géologique Du Maroc Au 1/50 000 Feuille De Skhour Des Rehamna* (2002)

35. A. A. Mouhsine, Evolution tectonothermale du massif hercynien des Rehamna (zone centre-mesetienne, Maroc) To cite this version: HAL Id: tel-01152558 (1994)

36. P. Jenny, *Contribution à la géologie structurale des Rehamna (Meseta marocaine méridionale). Le matériel paléozoïque et son évolution hercynienne dans le centre du massif.* Strasbourg, (1974)

37. European Space Agency, Sentinel-2 User Handbook. Issue 1 Revision 2, **48**(9) (1956)

38. M. Immitzer, F. Vuolo, C. Atzberger, First experience with Sentinel-2 data for crop and tree species classifications in central Europe. Remote Sens. **8**(3) (2016)

39. R. Chastain, I. Housman, J. Goldstein, M. Finco, Empirical cross sensor comparison of Sentinel-2A and 2B MSI, Landsat-8 OLI, and Landsat-7 ETM+ top of atmosphere spectral characteristics over the conterminous United States. Remote Sens. Environ. **221**, 274–285 (2019)

40. M. Drusch et al., Sentinel-2: ESA's optical high-resolution mission for GMES operational services. Remote Sens. Environ. **120**, 25–36 (2012)

41. L. Yan, D. P. Roy, H. Zhang, J. Li, H. Huang, An automated approach for sub-pixel registration of Landsat-8 Operational Land Imager (OLI) and Sentinel-2 Multi Spectral Instrument (MSI) imagery. Remote Sens. **8**(6) (2016)

42. A. Sekertekin, A. M. Marangoz, H. Akcin, Pixel-based classification analysis of land use land cover using Sentinel-2 and Landsat-8 data. Int. Arch. Photogramm. Remote Sens. Spat. Inf. Sci. ISPRS Arch. **42**(4W6), 91–93 (2017)

43. A. Novelli, M.A. Aguilar, A. Nemmaoui, F.J. Aguilar, E. Tarantino, Performance evaluation of object based greenhouse detection from Sentinel-2 MSI and Landsat 8 OLI data: A case study from Almería (Spain). Int. J. Appl. Earth Obs. Geoinf. **52**, 403–411 (2016)

44. G. Kaplan, U. Avdan, Object-based water body extraction model using Sentinel-2 satellite imagery. Eur. J. Remote Sens. **50**(1), 137–143 (2017)

45. K. N. Priyadarshini, V. Sivashankari, S. Shekhar, K. Balasubramani, Assessment on the potential of multispectral and hyperspectral datasets for land use/land cover classification. Proceedings **24**(1), 12 (2019)

46. J. M. Rodríguez Alves, J. M. P. Nascimento, J. M. Bioucas-Dias, A. Plaza, V. Silva, Parallel sparse unmixing of hyperspectral data. Int. Geosci. Remote Sens. Symp. **49**(6), 1446–1449 (2013)

47. A. Plaza, P. Martinez, J. Plaza, R. Perez, Spatial/spectral analysis of hyperspectral image data, in *2003 IEEE Work. Adv. Tech. Anal. Remote. Sensed Data*, vol. 00, no. C (2004), pp. 298–307

48. L. Genç, S. Smith, Assessment of principal component analysis (pca) for moderate and high-resolution satellite data. Trak. Univ J. Sci. **6**(2), 29–48 (2005)

49. X. Zhang, M. Pazner, N. Duke, Lithologic and mineral information extraction for gold exploration using ASTER data in the south Chocolate Mountains (California). ISPRS J. Photogramm. Remote Sens. **62**(4), 271–282 (2007)

50. A.B. Pour, M. Hashim, Identification of hydrothermal alteration minerals for exploring of porphyry copper deposit using ASTER data, SE Iran. J. Asian Earth Sci. **42**(6), 1309–1323 (2011)

51. S. Gabr, A. Ghulam, T. Kusky, Detecting areas of high-potential gold mineralization using ASTER data. Ore Geol. Rev. **38**(1–2), 59–69 (2010)

52. R. Amer, T. Kusky, A. El Mezayen, Remote sensing detection of gold related alteration zones in Um Rus area, central eastern desert of Egypt. Adv. Sp. Res. **49**(1), 121–134 (2012)

53. R. Amer, T. Kusky, A. Ghulam, Lithological mapping in the Central Eastern Desert of Egypt using ASTER data. J. African Earth Sci. **56**(2–3), 75–82 (2010)

54. A. P. Crósta, C. R. De Souza Filho, F. Azevedo, C. Brodie, Targeting key alteration minerals in epithermal deposits in Patagonia, Argentina, using ASTER imagery and principal component analysis. Int. J. Remote Sens. **24**(21), 4233–4240 (2003)

55. J. Zabalza, J. Ren, Z. Wang, H. Zhao, J. Wang, S. Marshall, Fast implementation of singular spectrum analysis for effective feature extraction in hyperspectral imaging. IEEE J. Sel. Top. Appl. Earth Obs. Remote Sens. **8**(6), 2845–2853 (2015)

56. A.A. Green, M. Berman, P. Switzer, M.D. Craig, A Transformation for ordering multispectral data in terms of image quality with implications for noise removal. IEEE Trans. Geosci. Remote Sens. **26**(1), 65–74 (1988)

57. J.B. Lee, A.S. Woodyatt, M. Berman, Enhancement of high spectral resolution remote-sensing data by a noise-adjusted principal components transform. IEEE Trans. Geosci. Remote Sens. **28**(3), 295–304 (1990)

58. G. Luo, G. Chen, L. Tian, K. Qin, S.E. Qian, Minimum noise fraction versus principal component analysis as a preprocessing step for hyperspectral imagery denoising. Can. J. Remote Sens. **42** (2), 106–116 (2016)

59. H.R. Matinfar, F. Sarmadian, S.K. AlaviPanah, R.J. Heck, Comparisons of object-oriented and pixel-based classification of land use/land cover types based on Lansadsat7, Etm+ spectral bands (case study: arid region of Iran). Am. J. Agric. Environ. Sci. **2**(4), 448–456 (2007)

60. M.J. Cracknell, A.M. Reading, Geological mapping using remote sensing data: a comparison of five machine learning algorithms, their response to variations in the spatial distribution of training data and the use of explicit spatial information. Comput. Geosci. **63**, 22–33 (2014)

61. A. S. Harvey, G. Fotopoulos, Geological mapping using machine learning algorithms. Int. Arch. Photogramm. Remote Sens. Spat. Inf. Sci. - ISPRS Arch. **41**, 423–430 (2016)

62. L. Breiman, Random forests. Random For. 1–122 (2001)

63. M. Pal, Random forest classifier for remote sensing classification. Int. J. Remote Sens. **26**(1), 217–222 (2005)

64. P.O. Gislason, J.A. Benediktsson, J.R. Sveinsson, Random forests for land cover classification. Pattern Recognit. Lett. **27**(4), 294–300 (2006)

65. B. Waske, J.A. Benediktsson, K. Arnason, J.R. Sveinsson, Mapping of hyperspectral AVIRIS data using machine-learning algorithms. Can. J. Remote Sens. **35**, S106–S116 (2009)

66. A.A. Othman, R. Gloaguen, Improving lithological mapping by SVM classification of spectral and morphological features: the discovery of a new chromite body in the Mawat ophiolite complex (Kurdistan, NE Iraq). Remote Sens. **6**(8), 6867–6896 (2014)

67. P. Duda, R. Hart, Pattern classification and scene analysis. **7**(4) (1973)

68. J. Kim, B.-S. Kim, S. Savarese, Comparing image classification methods: K-nearest-neighbor and support-vector-machines. Appl. Math. Electr. Comput. Eng. 133–138 (2012)

69. J. F. E. IV, D. Michie, D. J. Spiegelhalter, C. C. Taylor, Machine learning, neural, and statistical classification. J. Am. Stat. Assoc. **91**(433), 436 (1996)

70. R. Konieczny, R. Idczak, Mössbauer study of Fe-Re alloys prepared by mechanical alloying. Hyperfine Interact. **237**(1), 1–8 (2016)

71. P. D. Sugiyono, Data mining practical machine learning tools and techniques. **53**(9) (2016)

72. T. Hastie, R. Tibshirani, Discriminant adaptive nearest neighbor classification. IEEE Trans. Pattern Anal. Mach. Intell. **18**(6), 607–616 (1996)

73. G.J. Hay, T. Blaschke, Special issue: geographic object-based image analysis (GEOBIA). Photogramm. Eng. Remote Sens. **76**(2), 121–122 (2010)

74. B. Dezs et al., Object-based image analysis in remote sensing applications using various segmentation techniques. Ann. Univ. Sci. Budapest. Sect. Comp. **37**, 103–120 (2012)

75. A. Rekik, M. Zribi, A. Ben Hamida, M. Benjelloun, Review of satellite image segmentation for an optimal fusion system based on the edge and region approaches. Int. J. Comput. Sci. Netw. Secur. **7**(10), 242–250 (2007)

76. A. Baatz, M. Schäpe, Multiresolution segmentation: an optimization approach for high quality multi-scale image segmentation. **XII** (58) (2000)

77. P. Happ, R. Ferreira, C. Bentes, Multiresolution segmentation: a parallel approach for high resolution image segmentation in multicore architectures, in *The International Archives of the Photogrammetry, Remote Sensing and Spatial Information Sciences* (ITC, Enshede, 2010)

78. R.G. Congalton, A review of assessing the accuracy of classifications of remotely sensed data. Remote Sens. Environ. **37**(1), 35–46 (1991)

Optimization of Object-Based Image Analysis with Genetic Programming to Generate Explicit Knowledge from WorldView-2 Data for Urban Mapping

Azmi Rida, Amar Hicham, and Norelyaqine Abderrahim

Abstract

Object-based image analysis techniques give accurate results when a good knowledge base is extracted from remote sensing imagery. Data mining algorithms, especially the evolutionary process, can extract useful knowledge that can be used in different fields. In this paper, object-oriented classification was used, more particularly, the object-based image analysis approach (OBIA) is used to classify a large feature space composed of a very high spatial resolution (VHR) satellite image. The genetic programming (GP) concept was applied to extract classification rules with an induction form. This study aims to examine how data mining techniques based on the GP method can help to discover knowledge and extract classification rules automatically to illustrate well this knowledge. These rules are expected to enrich an anthology in the urban remote sensing domain. A comparison of the performance of three GP algorithms (Bojarczuk_GP, Falco_GP, and Tan_GP) was made using the JCLEC framework. Results showed two main conclusions. The first showed that generated rules can classify and extract useful knowledge from VHR satellite data using GP algorithms. The second demonstrates that the Bojarczuk model is efficient on accuracy classification than the Falco and Tan models.

Keywords

Remote sensing • High resolution • Data mining • Genetic programming • Rule-based system

1 Introduction

Knowledge-based systems (KBS) are becoming more and more important in various domains, especially in high-dimensional feature space where information is variable, and knowledge in this context is still complex to produce [1]. Indeed, acquiring and representing knowledge is a tedious process and the multiple steps involved in their creation can be very different according to the studied domain. This heterogeneity led to multiple questions and propositions, and the expert is often lost when the time comes to choose a solution. However, the advantages of representing and storing domain knowledge are undeniable. Indeed, it is then possible to produce intelligent systems based on the use of the acquired knowledge and to better explain and understand the domain under consideration.

In remote sensing, domain knowledge extraction is a tedious task. This is due to the complexity of the feature space, which is generally a satellite image with multiple spectral bands. The 1980s saw the emergence of satellites capable of producing high-resolution (HR) images between 30 and 10 m (Landsat-4, 1982; SPOT3, 1993). However, the 2000s appeared very high spatial resolution (VHR) satellite images whose spatial resolution is less than 5 m (QuickBird, 2001; PLEADEES, 2011). VHR satellites currently make it possible to obtain images with a resolution up to 0.5 m per pixel on the panchromatic band. Therefore, these images offer a much higher level of detail than HR images.

A new era has come to advance the semi-automatic extraction of objects from digital images. In the remote sensing field, multispectral imagery (MSI) captures reflected radiation over a series of adjoining bands, covering a very

A. Rida (✉)
Center of Urban Systems, Mohammed VI Polytechnic University (UM6P), 43150 Ben Guerir, Morocco
e-mail: rida.azmi@um6p.ma

A. Hicham
Mining Environment & Circular Economy, Mohammed VI Polytechnic University, 43150 Ben Guerir, Morocco

N. Abderrahim
Laboratoire de géophysique appliquée, de géotechnique, de géologie de l'ingénieur et de l'environnement (L3GIE), Mohammed V University – Mohammadia School of Engineering, Rabat, Morocco

© The Author(s), under exclusive license to Springer Nature Switzerland AG 2022
F. Barramou et al. (eds.), *Geospatial Intelligence*, Advances in Science, Technology & Innovation,
https://doi.org/10.1007/978-3-030-80458-9_12

large range of the electromagnetic spectrum for every pixel in the image. In the last decade, a new series of high spatial and spectral resolution imagery has become accessible and used more in different fields. Such images with sub-metric spatial resolution can provide features pertinent to the classification task, by enhancing accuracy and reducing spectral confusion in some cases.

However, the classification methods used with high spatial and spectral resolution data apply a new analysis technique called object-oriented image analysis approach or object-based image analysis approach (OBIA). These techniques are usually based on the use of domain knowledge [2]. The key issue in this approach is the obtainment of this knowledge, which is usually implicit and not formalized. Analysis methods must reduce the dimensionality of this very high-dimensional feature space to make any classification analysis more accurate [3, 4]. HR and VHR images have been increasingly used for the classification of land use land cover (LULC), but the spectral variation within the same class, the spectral confusion between the different land covers, and the shadow problem make per-pixel classifiers less efficient. The object-oriented classification approach is designed to deal with the problem of heterogeneity of the environment; it no longer treats the pixel in isolation but a group of pixels (objects) in their context [5].

The key parameter of the OBIA approach is the extraction of primitive objects from raw images, where each object corresponds to a group of homogeneous pixels. To recognize objects (or using methods able to detect objects), several techniques are generally based on the use of knowledge related to spectral, spatial, and contextual properties (e.g., spectral and textural values of an object, shape, length, area, form factor, etc.) [6].

About a decade ago, came the launch of the first software package specializing in OBIA: a revolutionary development in the remote sensing world that led to improvements within a wide field of applications. Over the past few years more packages have been developed, both specialized, and modules of existing image-analysis software.

A brief literature search reveals that publications in the early period of OBIA (2000 to 2003/04) were dominated by conference proceedings and "grey literature" but increasing numbers of empirical studies published in peer-reviewed journals have subsequently provided sufficient proof of the improvements that OBIA offers over per-pixel analyses. Figure 1 shows the increasing number of peer-reviewed articles published, and the number was doubled between 2006–2008.

The dimensions of the features extracted from image objects are much larger than pixels, which mainly contain spectral-based information (e.g., mean, ratio, and standard deviation). In object-based classification, hundreds of features involving the spectral, geometry, and texture features

can be obtained from the image objects. However, large amounts of features participating in classification always give rise to the "complexity of dimensionality", which decreases the classification accuracy. As some features make contributions to the classification and others have less influence on the result, features are commonly divided into relevant features, redundant features, and irrelevant features [2].

To yield better classification results, the irrelevant information should be removed, as much as possible, and the utilization of relevant information should be maximized. Therefore, feature selection before the object-based classification of high-resolution remote sensing images is a prerequisite. After the redundant and irrelevant features are removed, the training time is reduced, and the classification efficiency can be improved [7].

In the literature, only a few works focus on the development of a knowledge base to identify objects from remote sensing data. However, building a knowledge base in this context is not an easy task since the information required is generally variant and not formalized. This paper is organized into three sections as follows. In Sect. 2, the principles of knowledge extraction from remote sensing data and its relationship with GP were presented, as well as the algorithms used to realize this study. The methodology and the experiments were detailed in Sect. 3. Finally, Sect. 4 discussed the results and presented the concluding remarks.

2 Genetic Programming

Data mining technologies, e.g., fuzzy classifications [8], object-oriented classification (based on multiresolution segmented data) [7], per-pixel maximum likelihood [9], or artificial neural networks [10], have been used in several studies as a supervised or unsupervised remote sensing classification technique [11, 12]. However, using data characterized by huge volumes, high dimensionality, and having spatial attributes will be a tedious task capable of giving a result attended to be a suite. And of highly complex, high-dimensional, diversified, and variant datasets that present significant analysis challenges solving a problem automatically has always been the main interest. It was an idea that began in the late 1940s [13]. The domain of intelligent systems has always aimed at producing systems with supposedly intelligent behavior. GP is inspired by the design of natural evolution and seeks to solve problems automatically. An approach that requires intelligence if the same task is accomplished by a human being, is none other than the definition given by Arthur Samuel [13] on the purpose of automatic learning and intelligent systems. GP is a method inspired by the theory of evolution as it has been defined by Darwin [14], in particular its biological mechanisms. It aims to find programs that best meet a specified task. However,

Fig. 1 Cumulative number of publications using OBIA approach between 1990 and 2020 —scopus® database

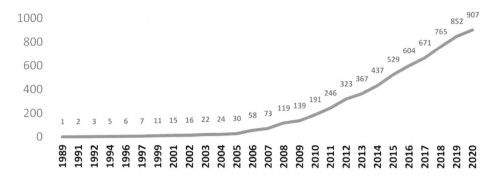

the GP concept allows the machine to learn, using an evolutionary approach, to optimize the programs' population.

Within the framework of GP from the first population of stochastically generated programs and using operators inspired by Darwinism, the GP evolves this population in a stochastic way. By reiterating this process, it is hoped to make the population converge toward solutions (programs) that respond to the problem to be solved. The flowchart showed in Fig.2 gives an idea of the general functioning of GP.

This diagram represents the operating cycle of a genetic program. First, a base is implemented to be able to start generating programs (initialization phase). Then, several individuals that generate future generations are obtained. At this time, a check is made to see if one of the solutions offered by these individuals is satisfactory (Evaluation Block). If no solution is suitable, a selection of the best must be made to generate descendants using different techniques like selection phase and crossover/mutation. Finally, these descendants will come to replace the previous generation by being, in turn, the parents, and the cycle then begins again with the evaluation block. Roughly, in biology, the

information carried by a gene is called a genotype, and the character expressed by this gene is called a phenotype [15]. By transposition in GP, a program can be seen from two angles: genotypic, the form on which the genetic operators apply, and phenotypic, the form in which the objective function or fitness function will be evaluated. The most common genotypic form in GP is the tree form, where each program is encoded as a tree. Reference [16] used this form to implement programs; it is the direct transposition of the prefixed form, used for example by the Lisp language [16].

This paper is motivated by the works of [17, 18] in analyzing and presenting the data structure of remote sensing data as a knowledge base to extract useful classification rules.

Finding a solid technique to extract knowledge from a feature space (VHR images) has two advantages: (i) being intuitively comprehensible to the user and (ii) being easily interpretable by problem-domain experts.

The induction form is one of the powerful techniques used in data mining techniques. Applying a rule-based system using the statement IF (conditions) THEN (predicts

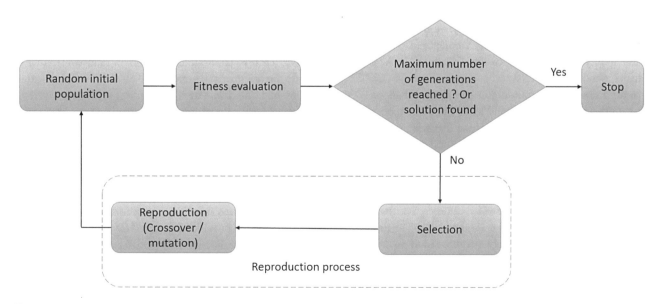

Fig. 2 General flowchart of genetic programming concept

class) is the challenge in this work. In the literature, there are several rule induction algorithms to discover such classification rules [19, 20]. A particularly famous strategy in computer science consists of the sequential covering approach, where in the essence the algorithm discovers one rule at a time until (almost) all examples are covered by the discovered rules (i.e., match the conditions of at least one rule). In contrast, sequential covering rule induction algorithms are mostly greedy, and they can perform a local search in the rule space.

An alternative approach to discover classification rules consists of using an evolutionary algorithm (EA), which performs a more global search in the rule space. Indeed, there are many EAs for discovering a set of classification rules from a given dataset [212223–24].

3 Methodology and Experiments

All data mining tasks involve at least three steps: (1) data preparation, (2) data analysis, and (3) decision-making. This work consists of the three fundamental steps listed as follows: first, the input data were preprocessed and prepared to extract knowledge from an imagery of WorldView-2 satellite sensor taken in 2011 [25]. The new knowledge base was analyzed to identify the relation between attributes and reduce spectral confusion in the dataset, and finally, GP was integrated as an optimization technique capable to find new and innovative classification rules. Figure 3 shows the complete processing chain for the proposed classification approach.

3.1 Study Area

The study area is Rabat city, the political capital of Morocco, located in the north-west of Morocco. Administratively, its territory has an area of $118.5 \ km^2$, composed of the urban municipality of Rabat, divided into five districts. At the last census conducted in 2014, its population was 5,77,827, making Rabat the seventh-largest city in the kingdom. With its suburbs, it forms the second-largest agglomeration of the country after Casablanca. Since June 2012, a group of city sites is inscribed on the UNESCO World Heritage List as cultural property. The heart of Rabat city is made up of the old city, to the west, and along the seafront, there is a succession of modern neighborhoods, and to the east, along the Bouregreg river. Between these two axes, going from north to south, there are three main neighborhoods: The first oneis Agdal, which is a very lively neighborhood of buildings mixing residential and commercial functions, mostly intended for the middle classes. The second is Hay Riad, the neighborhood with high-class areas that have experienced a

surge of dynamism since the 2000s, tending to become the new business center of Rabat. The last one is the Souissi neighborhood, consisting mainly of residential areas.

Hay Riad neighborhood made up of high standing houses with modern architecture was the study area of this work, where the roads are very clear, and the streets are also well visible. Rooftops have a unified geometry as well as their density allows a good segmentation of an input image.

3.2 Preprocessing of Input Data

The input data is generated mainly through a WorldView-2 satellite image which has eight multispectral bands: four (4) standard colors (red, green, blue, and near-infrared 1) and four (4) new bands (coastal, yellow, red edge, and near-infrared 2) [26]. WorldView-2 products are available as part of the DigitalGlobe Standard Satellite Imagery products from the QuickBird, WorldView-1/-2/-3, and GeoEye-1 satellites [27]. With the additional four spectral bands, WorldView-2 offers unique opportunities for remote sensing analysis of vegetation, coastal environments, agriculture, geology, and many other fields. This satellite image is characterized by high spatial resolution with 4 m for multi-spectral (MS) bands, and 0.5 m for the panchromatic (PAN) one. With its enhanced agility, WorldView-2 is capable of acting like a paintbrush, sweeping back and forth to collect very large areas of multispectral imagery in a single pass. The sensor can collect nearly 1 million km^2 every day; its high altitude allows it to typically revisit any place on earth in 1.1 days.

Radiometric calibration

As a preprocessing step, a radiometric correction was used to prepare the data for segmentation and extraction of the knowledge base. Radiometric correction of MS and PAN data was used to calibrate aberrations in data values due to specific distortions from atmosphere effects (such as haze) or instrumentation errors (such as striping) [21].

DigitalGlobe sensor products (image pixels) are radiometrically corrected image pixels. Their values are a function of how much spectral radiance enters the telescope aperture and the instrument conversion of that radiation into a digital signal [28]. Therefore, image pixel data are unique to each sensor.

A calibration step has been performed (at provider level) and these data are provided in the *.IMD* metadata file that is delivered with the imagery. Since its launch, DigitalGlobe performs an extensive vicarious calibration campaign to provide an adjustment to the prelaunch values. The top of atmosphere radiance (L) in units of $[W \mu m^{-1} \ m^{-2} \ sr^{-1}]$ is

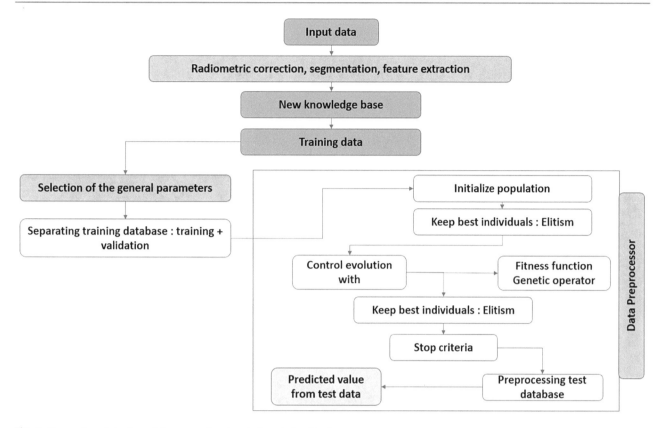

Fig. 3 Processing chain from data preparation to rule-based classification

then calculated for each band by converting from digital numbers (DN).

Equation (1) is used to convert the at-sensor radiance to top of atmosphere reflectance where calculations are performed independently for each band and pixel:

$$\rho_{\lambda_{pixelBand}} = \frac{L_{\lambda_{pixelBand}} * d_{ES}2 * \pi}{E_{sun\lambda_{Band}} * \cos(\Theta)_s} \qquad (1)$$

- "L" is at-sensor radiance calculated from data provided in .IMD file;
- "d_{ES}" is the Earth–Sun distance in astronomic unit;
- E_{sun} is the band-averaged solar exoatmospheric irradiance;
- Θ is the solar zenith angle (90-meanSunEl from IMD file).

Pan-sharpening

The second preprocessing step is the pan-sharpening of the data, where this technique was used to enhance spatial resolution. Recently, several applications, such as land-cover classification, feature extraction, image segmentation, and change detection, require both spatial and spectral images for fine features detection in suburban or urban scenes. The literature shows a large collection of pan-sharpening

methods developed and used to enhance spatial resolution and preserve spectral information. In this study, *NNDeffuse* algorithm developed by [29] was used to fusion MSI and PAN data.

Segmentation

Segmentation is a main preprocessing step that allows the user to identify the object that has similar spectral characteristics pixels. It is the process of completely portioning a scene (in this case remote sensing image) into non-overlapping regions (segments) in scene space. In the segmentation process, all objects are outlined without any class label. Usually, the outlined objects should have one specific object, to generate appropriate segments capable to distinguish semantic objects (Fig. 4).

Many powerful algorithms have been developed within pattern recognition and computer vision since the 1980s, where research led to successful applications in disciplines like medicines or telecommunication engineering. However, their application in the fields of remote sensing and photogrammetry was limited to special purpose implementations only. Nevertheless, this limitation is due to the complexity of the underlying object models and the heterogeneity of sensor data in use. With the appearance of high spatial resolution

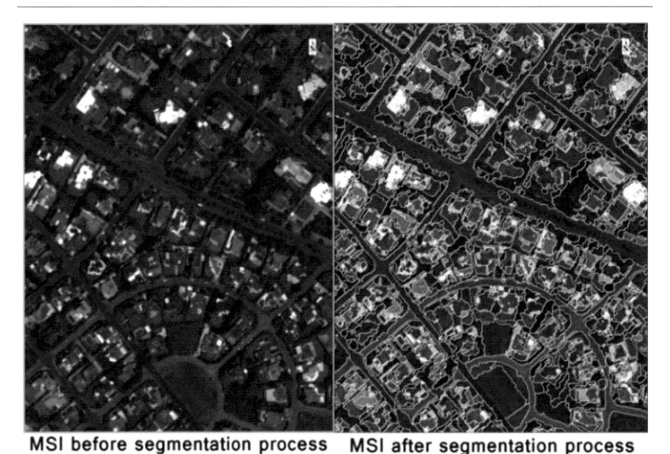

MSI before segmentation process   MSI after segmentation process

Fig. 4 Example of Hay Ryad district (in Rabat city—Morocco) showing WorldView-2 image before and after segmentation process

satellite data as well as multisource data sources, the segmentation methods have become evident again, and significant progress has been made with the introduction of the first commercial and operational software product (eCognition by Definiens-Imaging) in 2000 [30].

Segmentation methods follow two strongly correlated principles of neighborhood and value similarity. A watershed algorithm (WA) for segmentation is used. This method integrates duplicate neighboring areas based on a combination of spectral and spatial information. The WA transform is based on the concept of hydrologic watersheds, where basins fill up with water starting at the lowest points, and dams are built and water coming from different basins would meet. The process stops when the water level has reached the highest peak in the landscape [31]. A similar process is applied in digital imagery using the luminosity of the pixel; the darker the pixel, the lower the elevation. A watershed algorithm sorts pixels by increasing the grayscale value and then begins with the minimum pixels and "floods" the image, partitioning the image into regions with similar pixel intensities based on the computed watersheds. The result is a segmented image, where each region is assigned to the mean spectral values of all the pixels that belong to that region.

In this study, the Full Lambda-Schedule algorithm developed by [32] is used to merge segments. The algorithm iteratively merges adjacent segments based on a combination of spectral and spatial information as mentioned above. Figure 4 shows results after segmentation over Hay Riad district where individual buildings are surrounded as well as the green spaces (of grass and trees) and road networks.

3.3 Feature Extraction

All segmented objects from VHR images have spectral, spatial, and textural features, to have an accurate classification process. More than one attribute characterizing an object must be found to explain this accurate classification; for instance, shadow class has a high spectral value in near-infrared bands, and grass has a high rectangularity index in urban areas with a coarse texture and mean NDVI values. Combining several distinctive attributes for each class will facilitate the extraction of useful classification rules.

After the segmentation process, attributes of each segment were calculated and extracted using a feature extraction module implemented in ENVI 5.0 software [33]. Attributes were categorized as spatial, spectral, and textural attributes. Additional data was calculated using a normalized band ratio (infrared and red) and calculation of hue, saturation, and intensity (HSI) attributes. The database extracted from ENVI's module was a *dbase file composed of 111 attributes. Details about calculated attributes can be found in [34]. A number of 590 segments as a training set was used, where the proposed classes are shown as follows (Shadow, Built up_Roofs, Built Up_Roads, Vegetation_Grass, Vegetation_Trees, Bare soil, and Water).

In Fig. 4, there are three levels of object classes. The first level contains the main component of the urban ecosystem (Built up, Water, and vegetation). In the case of VHR image, the shadow class was added due to the high buildings and trees. The second level of image object contains derived information from the three components of the first-class object. In the last level (3), there are more details about one specific class or sub-classes (Fig. 5).

Table 1 resumes all calculated and extracted attributes from a segmented image. All attributes that were in the rule-based system are implemented in ENVI software to classify the input image. Attributes were divided into two types of bands (spectral and derived bands), respectively, spectral indices and calculated attributes such as spectral, geometric, and textural attributes.

3.4 Feature Selection

In modern machine learning algorithms, there are methods used to reduce dimensionality [35, 36]. In general, these tasks are rarely performed in isolation. Instead, they are often preprocessing steps to support other tasks. In literature,

there are two main strategies of dimension reduction: (i) Feature selection techniques that are typically grouped into three approaches, namely filter, embedded, and wrapper methods that extract subsets from existing features, and (ii) feature extraction (e.g., principal component analysis—PCA) [37]. The key difference between feature selection and extraction is that feature selection keeps a subset of the original features while feature extraction creates brand new ones.

In this paper, the selection of attributes was made for the supervised classification. In this context, the objective of selection is finding an optimal subset of attributes that can be composed of relevant attributes and must seek to avoid redundant ones. In addition, this set must make it possible to best meet the objective set, namely the accuracy of learning, the speed of learning, or even the applicability of the proposed classifier.

ReliefF-based feature selection method was used in this paper, where it takes a filter method approach [38]. The proposed method was used to calculate a feature score for each feature. This score can be applied to rank and select top-scoring features. Many researchers adopted the ReliefF algorithm to preliminary filter high-dimensional features in the feature database [38, 39].

By applying the ReliefF method on the input dataset, results made it possible to select the 20 best attributes of which it proposed bands 6, 7, and 8, which are, respectively, red edge, near-infrared—NIR-1, and near-infrared—NIR-2. Also, the hue, saturation, and intensity (HSI) transformation from RGB bands were highly ranked and used in the new filtered dataset.

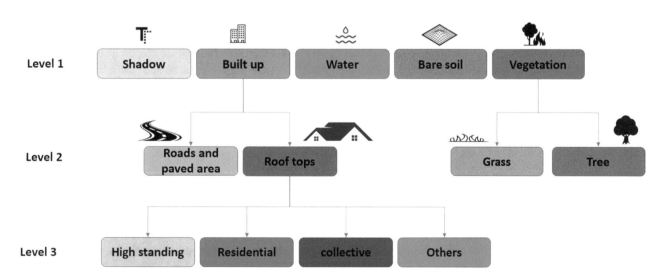

Fig. 5 Object classes hierarchy used in this study

Table 1 Extracted parameters from preprocessed satellite image

Attributes	Class attribute
Spectral bands	Spectral information was calculated using the 8 input bands where (Min, Max, Mean, and Standard Deviation was calculated)
Derived bands	Hue—Saturation—Intensity
Spectral Indices	Band ratio using Red and NIR bands
Geometric calculation	12 variables are calculated
Textural calculation	4 variables are calculated

3.5 Generating Classification Rules

In this paper, knowledge is presented as multiple IF–THEN rules in the decision rules list. Such rules state that the presence of one or more conditions (antecedents) implies or predicts the presence of other conditions (consequents). A typical rule has the form of: If X1 and X2 and … Xn as conditions THEN Y, where Xi $\in$ (1, 2,..,n) is the antecedent that leads to the prediction of consequent Y. The reason why classification rules were used instead of the decision tree is because each rule can be seen as an independent piece of knowledge. Thus, newly generated rules can be added to an existing ruleset without disturbing existing ones. Multiple rules can be concatenated to form a set of decision rules. This last is usually listed according to their accuracy, where the best rule was listed first.

The Java Class Library for Evolutionary Computation— JCLEC Framework developed in Java Environment was used. JCLEC is a representative example of an evolutionary optimization framework designed for one main objective, to maximize its reusability and adaptability to new paradigms with a minimum of programming effort [40, 41]. The implemented classification module is an intuitive, usable, and extensible open-source module for GP classification algorithms [42]. This module is a part of open-source software for researchers and end-users to develop and use classification algorithms based on GP and grammar-guided genetic programming (3GP) models [43], an extension of GP which makes the knowledge extracted more expressive and flexible using a context-free grammar.

JCLEC classification module houses three 3GP classification algorithms listed as follows: (Bojarczuk_*GP* [44], Falco_GP [21] and Tan_GP [45]). JCLEC extends a class called PopulationAlgorithm. This parent class defines the main steps of the evolutionary process. To initialize the population, a component is triggered with the number of solutions to be created as a parameter; in this case the number of solutions is equal to the n class in the dataset. Each solution individual should contain a fitness object representing its quality.

Bojarczuc Model

The author used GP standard operators to evolve decision trees using a defined syntax. Bojarczuk used a GP-based approach, where a set of functions applicable to different types of attributes is defined to represent the rules as a disjunctive normal form. Several constraints are placed on the tree structure to express a valid rule. This type of GP is also referred to as constrained syntax GP [44, 46]. The fitness function used in the Bojarczuk model evaluates the quality of each individual (a rule set where all rules predict the same class) according to two basic criteria, namely its predictive accuracy and its simplicity [47]. Implementation of this fitness function in the JCLEC module is a subclass called BojarczukEvaluator. The fitness function in this case evaluates the confusion matrix for each of the data classes [44].

Falco Model

The author used GP to evolve comprehensible simple rules by combining the parallel searching ability of genetic programming. Falco used a classifier tree that is constructed using logical functions and attribute values. A grammar has been designed that can represent such rules. The author has shown that the evolved rules are comprehensible, emphasize discriminating variables, and achieve compatible performance as compared to other classification algorithms on benchmark datasets [21]. The fitness function used in this case evaluates the number of prediction errors for the class of the current algorithm's execution [21].

Tan Model

Tan model is based upon a modified version of steady-state GP in [48]. The fitness function evaluates the quality of each rule or individual, which is based on the evaluation function defined in Eq. (2). In other words, the fitness function evaluates the confusion matrix for the data class of the current algorithm's execution.

$$Fitness = \frac{Tp}{Tp + W1 * Fn} * \frac{Tn}{Tn + W2 * Fp} \qquad (2)$$

where Tp, Fp, Tn, and Fn stand for true positive, false positive, true negative, and false negative, respectively. $W1$ and $W2$ are the weights; they enable the dependency of fitness function on different concepts.

Nevertheless, the performance of three modules has been tested and validated in the following section with statistical tests like R-squared, p-value, and ROC area.

4 Results and Discussion

GP models interpretation

The models used in this study generated explicit rules to simplify knowledge from high-dimensional feature space. Rule induction technique that creates the "If—Then—Else" type was used; it generates rules from a set of input variables, and it can work with both numerical and categorical values. In this case, an inductive prediction that concludes a future instance from a past sample is as follows:

```
IF (antecedents)₁ THEN class₁
ELSE IF (antecedents)₂ THEN class₂
ELSE default class
```

All the models proposed the same inductive structure with a variety of attributes that were candidates to generate an optimal solution and help the user in the choice of the most representative attributes in the search space. The interpretation of these choices is based on three main classes: shadow, roofs, and trees.

The rules below show an example for the result of Bojarczuk's model, presented in the inductive form. For shadow class, the algorithm proposed 2 bands, respectively, the blue and near-infrared band 2—(NIR-2). Dark objects, which confound many shadow detection algorithms, often have much higher reflectance in the NIR band. The blue band is also considered to be an excellent choice because several studies have shown that shadow pixels are illuminated by the predominantly blue, diffuse sky radiation. Bojarczuk model suggested a combination between NIR-2 and blue band where the real interpretation of the rule is *IF average value of blue band < = 19,188 AND the average value of NIR-2 band < = 37,71,600 THEN objects belong to shadow class*. On the other hand, Falco's model proposed the texture of band 6 (red-edge band) to represent the shadow class. The red-edge band is a division of the red spectrum between 700 and 750 μm. However, the red band

reflects a small part of the dark pixels which is considered a poor choice for this class. The Tan model suggested a complicated rule composed of four conditions and two logical operators ("AND" & "AND NOT"); the NIR-1 and NIR2 bands, the ratio between the red band and NIR (which is the ratio of the Normalized Difference Vegetation Index—NDVI). NDVI band should not be taken into account according to the Tan model using the logical operator (AND NOT), and finally the minimum value and the NIR-1 band. This complexity can cause interference between conditions.

```
IF (AND < = AVG_B2 191,880,007 < = AVG_B8 377,160,068)
THEN (Class = Shadow.
ELSE   IF   (AND < = AVG_B9   -0,608,389 > TXAVG_B7
430,725,437)
THEN (Class = vegetation_2)
ELSE   IF   (AND > AVG_B2   191,880,007 < = AVG_B8
377,160,068)
THEN (Class = water)
ELSE IF (AND < = MAX_B9 -0,317,148 AND < = TXAVG_B7
430,725,437 > MIN_B10 107,685,121)
THEN (Class = Vegetation_1)
ELSE IF (AND < = TXRAN_B11 0,083,049 AND < = TXAVG_B7
430,725,437 > MAX_B9 -0,317,148)
THEN (Class = Built_up_1)
ELSE IF (AND > AVG_B9 -0,608,389 > AVG_B8 377,160,068)
THEN (Class = Built_up)
ELSE IF (AND < = TXRAN_B11 0,083,049 AND < = MIN_B10
107,685,121 < = TXAVG_B7 430,725,437)
THEN (Class = Bare_Soil)
ELSE (Class = Built_up)
```

For buildings and rooftops, Bojarczuk's model proposed the average value of NDVI band and NIR-2 band. Buildings and rooftops have particular characteristics relative to other features. For example, the shape of rooftops approximates a rectangle, the area of rooftops of residential buildings is within a certain range, compared to industrial or other types of buildings. In our case, rooftops of interest are relatively dark, so they should have a low average pixel value (< 0.4). However, NDVI would be a good criterion to start with in this example, where the buildings have smaller NDVI values than vegetation.

The Falco model suggested the average value of the blue band and the minimum of the NIR-2 band. The near-infrared bands may contain low reflectance for dark pixels, which may meet our needs but not with great certainty. Dark pixels can also exist in the roads and bare ground classes. The Tan model once again proposed a rule with three conditions, but very interesting attributes. He suggested the minimum values for the NDVI band (which is a very good choice), the average texture values for NIR-1 band, and texture range for band 11 (the HSI transformation of RGB bands).

In the case of trees class, the Bojarczuk model proposed the maximum value of the NDVI, which was predicted, and the mean texture of the NIR-1 band and the minimum value of the band 10 (which is the derived HIS transformation from RGB bands). In the literature, it is known that trees are more textured than grass, so the choice of the mean texture within the infrared band is well done for the Bojarczuk model.

In the case of the Falco model, the algorithm proposed NDVI combined with the coastal blue band (band 1). This combination can lead to inaccuracy of the generated rule due to the coastal blue band reflectance values. Thus, the Tan model has proposed the minimum value of band 10.

4.1 Validation Metrics

As a validation method, the confusion matrix was used to evaluate classification accuracy, which is a common way of presenting true positive (TP), true negative (TN), false positive (FP), and false-negative (FN) predictions. Those values are presented in the form of a matrix where the Y-axis shows the true classes while the X-axis shows the predicted classes.

Table 2 shows that the Bojarczuk model's classification gives a diagonal matrix, except for the confusion between (roads and roofs) classes, a large part of the roofs has been classified as roads. This is generally due to the spectral properties of the infrared band where both classes contain dark pixels. Also, a spectral confusion between roofs and water areas has been provoked. This is due generally to the low reflectance of water pixels in some areas. A second confusion has been shown in Table 3 with the trees and grass classes. This confusion is due to the spectral rapprochement between trees and grass classes in the NDVI index.

The Falco and Tan models show the same confusion (roads and roofs) with a slight difference between the number of misclassified lines.

4.2 Statistical Metrics

The confusion matrix was presented in Table 3 to evaluate the behavior of the three models in terms of classification accuracy. On the other hand, it can be more flexible to predict the probabilities of an observation belonging to each class in a classification problem rather than predicting classes directly. This flexibility comes from the way that probabilities may be interpreted using different thresholds that allow the model to trade off concerns in the errors made by the model, such as the number of false positives compared to the number of false negatives. This is required using models where the cost of one error outweighs the cost of other types of errors.

ROC areas and precision-recall curves (PRC) were used to explain the probabilistic forecast for binary (two-class) classification predictive modeling problems [42]. The metrics are used to evaluate the performance of the three models. Precision can be understood as a measure of accuracy or quality, while recall is a measure of completeness or quantity [49].

A measure that combines precision and recall in their harmonic mean, called F-measure or F-score, is used to estimate model performance. However, one rule is used: the higher the score, the better the model. This parameter combines precision and recall into one metric. Table 3 shows that Bojarczuk has the best F-score for all classes, followed by Falco and Tan. Another measurement parameter that distinguishes the performance of several models is the ROC area. In general, ROC curves are based on the rate of true positives (TP Rates) and the rate of false positives (FP Rates). These are relationships that do not depend on the distribution of classes. This robust method eliminates the need to know the costs of classification and the distribution of classes. To calculate the points of a ROC curve, several evaluations of a logistic regression model are performed by varying the classification thresholds, but this would be ineffective. In other words, the AUC provides an aggregated measure of performance for all possible classification thresholds. AUC can be interpreted as a measure of probability for the model to classify a random positive example above a random negative example. Table 3 shows the AUC values for the three models for all classes. Bojarczuk's model showed again a great score compared to other models.

In general, a value of ROC-AUC greater than 0.7 is a good representative value for a model. However, Table 3 shows that the three models were able to classify the seven classes with a score beyond 0.7, except for the road and trees classes in the Falco model. It seemed, however, that the Falco model is unable to distinguish between these two classes precisely. A second anomaly is noticed between (bare soil) and (roads), where the Tan model found difficulty in classifying these two instances correctly.

It is highly recommended to use precision-recall curves as a supplement to the routinely used ROC curves to get the full picture when evaluating and comparing tests. It is used less frequently than ROC curves but as we shall see PRC may be a better choice since the current dataset contains imbalanced data. Since precision-recall curves do not consider true negatives, they should only be used when specificity is of no concern for the classifier. In other words, the PRC area represents a different trade-off which is between the true positive rate and the positive predictive value.

PRC is simply a graph with precision values on the y-axis and recall values on the x-axis. In other words, the PRC

Table 2 Confusion matrix for the Bojarczuk, Falco, and Tan models

	Confusion matrix	Bare soil	Grass	Road	Roofs	Shadow	Trees	Water
Bojaczuk model	Bare soil	42	2	9	10	0	3	0
	Grass	0	110	0	0	0	4	0
	Road	1	0	30	2	1	0	0
	Roofs	8	1	27	72	6	0	15
	Shadow	3	0	2	1	51	5	0
	Trees	11	18	3	4	8	61	0
	Water	0	1	0	1	0	0	70
Falco model	Bare soil	57	0	0	8	0	1	0
	Grass	16	98	0	0	0	0	0
	Road	25	0	1	4	0	0	4
	Roofs	46	4	0	76	1	0	2
	Shadow	32	0	0	0	26	4	0
	Trees	65	7	0	1	0	32	0
	Water	8	0	0	1	1	0	62
Tan model	Bare soil	19	1	15	17	7	6	1
	Grass	3	105	1	3	0	2	0
	Road	1	0	11	11	10	0	1
	Roofs	2	3	11	102	3	1	7
	Shadow	0	0	2	4	40	12	4
	Trees	4	13	1	1	14	72	0
	Water	1	1	1	0	0	0	69

Table 3 Evaluation parameters of the three models

–	Bojarczuk			Faclo			Tan		
	ROC Area	PRC Area	F-Score	ROC Area	PRC Area	F-Score	ROC Area	PRC Area	F-Score
Bare soil	0.80	0.45	0.64	0.75	0.21	0.36	0.64	0.24	0.40
Grass	0.95	0.81	0.90	0.91	0.80	0.88	0.95	0.80	0.88
Roads	0.90	0.38	0.57	0.52	0.08	0.057	0.64	0.12	0.28
Roofs	0.75	0.54	0.65	0.77	0.59	0.70	0.85	0.63	0.76
Shadow	0.90	0.65	0.80	0.70	0.45	0.57	0.80	0.39	0.58
Trees	0.78	0.56	0.68	0.64	0.39	0.45	0.82	0.59	0.72
Water	0.97	0.80	0.90	0.93	0.80	0.88	0.96	0.81	0.90

contains TP/(TP + FN) on the y-axis and TP/(TP + FP) on the x-axis. Both precision and recall are important metrics to evaluate the performance of the binary classification model. The corresponding PRC values in Table 3 show the loss of precision, even the ROC-AUC area of bare soil class that was 0.80 in the Bojarczuk model, barely touches on 0.45 precision. This deficit allows concluding that even higher ROC-AUC can hide a lot of imprecision in some cases. Falco and Tan models show much lower values in the PRC area.

Finally, a weighted average value of all metrics shows that the Bojarczuk model had high accuracy followed by Tan and *Falco*. The performance evaluation of the three models with the application of the AUC-ROC curve (specificity vs sensitivity) and PRC demonstrates that the Bojarczuk model is efficient than the Falco and Tan models.

There is a correlation between statistical metrics and attributes generated in the proposed rules. The model that performed well in terms of statistical attributes is the same model that proposed good rules. Based on developed

Fig. 6 Graph showing the accuracy chart for the three models based on statistical metrics

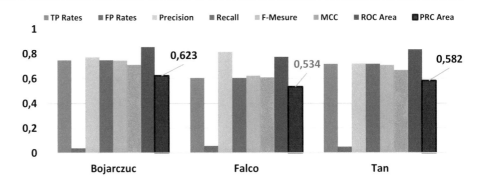

expertise in choosing the right attributes (spatial, spectral, textural, or even derived products such as NDVI), the interpretation of the rules generated by Bojarczuk's model shows a good choice of these attributes (Fig. 6).

In this paper, the evaluation of the models was limited at the level of their statistical metrics. Execution of the rule classification has shown that the models can extract each class separately from the other classes; in other words, the classification rules generated by the three models can make a good extraction of one class at a time. Tests performed in the ENVI software, using its feature extraction module, have shown that applying the rule to a single class is capable to improve accurately extracting the class.

5 Conclusion

The performance of the evolutionary approach was tested; particularly, genetic programming is used to extract explicit knowledge from VHR satellite images. Genetic programming algorithms have shown their performance in explicit knowledge extraction, especially in a complex feature space.

Genetic programming has shown its ability to simulate human expertise in the choice of the most representative variables to apply a rule-based supervised classification. Despite advances in the development of various proposed algorithmic models, the evolutionary approach is still unable to detect a precise threshold value for a given class.

However, a perspective can be retained from this work, focusing on strengthening the algorithmic model so that it can detect more accurate threshold values. This is feasible if a large amount of training data is given, as well as eliminating the preprocessing part that allows filtering of variables that have more influence on the feature space.

References

1. M.L. Brodie, J. Mylopoulos, *on Knowledge Base Management Systems: Integrating Artificial Intelligence and Database Technologies*. (Springer Science & Business Media, 2012)

2. L. Zhang et al., A knowledge-based procedure for remote sensing image classification, in *2014 11th International Conference on Fuzzy Systems and Knowledge Discovery (FSKD)*. (IEEE, 2014)

3. R. Azmi, O.B. Alami, A.E. Saadane, I. Kacimi, T. Chafiq, A modified and enhanced normalized built-up index usingmultispectral and thermal bands. Ind. J. Sci. Technol. **9**(28), art. no. 87405, 1-10 (2016). https://doi.org/10.17485/ijst/2016/v9i28/87405

4. L. Gu, Q. Cao, and R. Ren, Building extraction method based on the spectral index for high-resolution remotesensing images over urban areas. J. Appl. Remote Sens. **12**(4), 045501 (2018).https://doi.org/10.1117/1.JRS.12.045501

5. T. Blaschke, Object based image analysis for remote sensing. ISPRS J. Photogramm. Remote Sens. **65**(1), 2–16 (2010)

6. L.J. Quackenbush, A review of techniques for extracting linear features from imagery. Photogram. Eng. Remote Sens. **70**(12), 1383–1392 (2004)

7. M.R. Rahman, S.J. Saha, Multi-resolution segmentation for object-based classification and accuracy assessment of land use/land cover classification using remotely sensed data. J. Indian Soc. Remote. Sens. **36**(2), 189–201 (2008)

8. A. Amo et al., Fuzzy classification systems. **156**(2), 495–507 (2004)

9. P. Bolstad, T. Lillesand, R. Rapid maximum likelihood classification. Photogram. Eng. Remote Sens. **57**(1), 67–74 (1991)

10. Zurada, J.M., *Introduction to Artificial Neural Systems*, vol. 8. (West Publishing St. Paul., 1992)

11. J. Han, J. Pei, M. Kamber, *Data Mining: Concepts and Techniques*. (Elsevier, 2011)

12. R. Azmi, A. Saadane, I. Kacimi, and M. Hakdaoui, Hybrid classification approach combining objectoriented method and expert system for extracting land cover map from very high spatial resolution image – case study city ofRabat – Morocco. Int. J. Innovation Appl. Stud. **10**(2), 594–603 (2015)

13. C.J.M. Machinery, Computing machinery and intelligence-AM Turing. **59**(236), 433 (1950)

14. S.J.J.S. Gould, Darwinism and the expansion of evolutionary theory. **216**(4544), 380–387 (1982)

15. C.H. Waddington, *An Introduction to Modern Genetics*. (1939)

16. J.R. Koza, J.R. Koza, *Genetic Programming: on the Programming of Computers by Means of Natural Selection*, vol. 1. (MIT Press, 1992)

17. K. Mertens et al., Using genetic algorithms in sub-pixel mapping. **24**(21), 4241–4247 (2003)

18. D Sheeren et al., Discovering rules with genetic algorithms to classify urban remotely sensed data, in *Proceedings IEEE International Geoscience and Remote Sensing Symposium (IGARSS'2006)* (2006)

19. P. Clark, R. Boswell. Rule induction with CN2: some recent improvements, in *European Working Session on Learning*. (Springer, 1991)

20. J.W. Grzymala-Busse, *A new version of the rule induction system LERS*. Fandom. Inform. **31**(1), 27–39 (1997)

21. I.D. Falco, A.D. Cioppa, E. Tarantino, *Discovering interesting classification rules with genetic programming.* Appl. Soft Comput. **1**(4), 257–269 (2002)

22. M.V. Fidelis, H.S. Lopes, A.A. Freitas, *Discovering comprehensible classification rules with a genetic algorithm*, in *Proceedings of the 2000 Congress on Evolutionary Computation. CEC00 (Cat. No. 00TH8512).* (IEEE, 2000)

23. A.A. Freitas, A survey of evolutionary algorithms for data mining and knowledge discovery, in *Advances in evolutionary computing.* (Springer, 2003), pp. 819–845

24. P.J. Sharma, Discovery of classification rules using distributed genetic algorithm. Procedia Comput. Sci. **46**, 276–284 (2015)

25. M.A.C. Tarantino, G. Pasquariello, F. Lovergine, P.B. Valeria, Tomaselli, 8-Band Image Data Processing of the Worldview-2 Satellite in a Wide Area of Applications. Earth Observation. (2012)

26. R.P.S. Landry, A comparative analysis of high spatial resolution IKONOS and WorldView-2 imagery for mapping urban tree species. Remote Sens. Environ. **124**, 516–533 (2012)

27. T. Novack et al., Machine learning comparison between WorldView-2 and QuickBird-2-simulated imagery regarding object-based urban land cover classification. Remote Sens. **3**(10), 2263–2282 (2011)

28. C. Tarantino et al., 8-band image data processing of the worldview-2 satellite in a wide area of applications. *Earth Observation.* (2012) p. 137–152

29. W. Sun, B. Chen, D.W. Messinger, Nearest-neighbor diffusion-based pan-sharpening algorithm for spectral images. Opt. Eng. **53**(1), 013107–013107 (2014)

30. J. Tian, D.M. Chen, Optimization in multi-scale segmentation of high-resolution satellite images for artificial feature recognition. Int. J. Remote Sens. **28**(20), 4625–4644 (2007)

31. J. Roerdink, A. Meijster, The watershed transform: definitions, algorithms and parallelization strategies, Fundam. Inform. **41**(1,2) 187–228 (2001)

32. D.J. Robinson, N.J. Redding, D.J. Crisp, *Implementation of a Fast Algorithm for Segmenting SAR Imagery.* (Electronics Research Lab Salisbury, Australia, 2002)

33. ITT, *ENVI EX User's Guide.* (ENVI, 2009) p. 52–75

34. ENVI, *Feature Extraction with Example-Based Classification Tutoria.* (2010) http://www.harrisgeospatial.com/portals/0/pdfs/envi/FXExampleBasedTutorial.pdf

35. P.S. Bradley et al., *Method of Reducing Dimensionality of a Set of Attributes Used to Characterize a Sparse Data Set.* (Google Patents, 2004)

36. G. Chandrashekar, F.J.C. Sahin, E. Engineering, A survey on feature selection methods. **40**(1), 16–28 (2014)

37. R. Bro, A.K. Smilde, Principal component analysis. Anal. Methods **6**(9), 2812–2831 (2014)

38. K. Kira, L.A. Rendell, A practical approach to feature selection, in *Machine Learning Proceedings 1992.* (Elsevier, 1992), pp. 249–256

39. Z. Wang et al., Application of ReliefF algorithm to selecting feature sets for classification of high resolution remote sensing image. in *2016 IEEE International Geoscience and Remote Sensing Symposium (IGARSS).* (IEEE, 2016)

40. A. Cano et al., A classification module for genetic programming algorithms in JCLEC. **16**(1), 491–494 (2015)

41. JCLEC, *JCLEC Classification Module/User Manual and Developer Documentation.* (2014), http://jclec.sourceforge.net/data/JCLEC-classification.pdf

42. R. Azmi, H. Amar and A. Norelyaqine, Generate knowledge base from very high spatial resolution satellite image usingrobust classification rules and genetic programming. 2020 IEEE International conference of Moroccan Geomatics (Morgeo), 1–6 (2020). https://doi.org/10.1109/Morgeo49228.2020.9121914

43. R.I. Mckay et al., Grammar-based genetic programming: a survey. **11**(3–4), 365–396 (2010)

44. C.C. Bojarczuk et al., A constrained-syntax genetic programming system for discovering classification rules: application to medical data sets. **30**(1), 27–48 (2004)

45. K.C. Tan et al., Mining multiple comprehensible classification rules using genetic programming. in *Proceedings of the 2002 Congress on Evolutionary Computation. CEC'02 (Cat. No. 02TH8600).* (IEEE, 2002)

46. H. Jabeen, A.R. Baig, Review of classification using genetic programming. Int. J. I. Eng. Sci. Technol. **2**(2), 94–103 (2010)

47. C.C. Bojarczuka, H.S. Lopesb, A.A. Freitasc, Data mining with constrained-syntax genetic programming: applications in medical data set. **6**, 7–0 (2001)

48. W. Banzhaf et al., *Genetic Programming.* (Springer, 1998)

49. T. Saito, M. Rehmsmeier, The precision-recall plot is more informative than the ROC plot when evaluating binary classifiers on imbalanced datasets. Plos One **10**(3), e0118432 (2015)

Machine Learning and Remote Sensing in Mapping and Estimating Rosemary Cover Biomass

Hassan Chafik, Mohamed Berrada, Anass Legdou, Aouatif Amine, and Said Lahssini

Abstract

Biomass estimation is important to predict the production of medical and aromatic shrubs. This work presents an efficient method to estimate the biomass of rosemary cover based on satellite imagery data. The approach consists of using remote sensing and machine learning techniques to map the rosemary cover, then estimating the dry biomass using a simple regression model that estimates the weight of a single rosemary tuff. The results are maps of the rosemary cover density where the random forest classifier gave high validation scores (67.5, 75.5, and 80%). The tuff weight estimator gave an accuracy of 7%.

1 Introduction

Grasslands are areas where the vegetation is dominated by grasses. However, sedges and rushes are also found here, as well as other grasses in varying proportions. Grasslands occur naturally on every continent except Antarctica and are found in most ecoregions of the earth. Besides, grasslands are one of the largest biomes on the planet and dominate the landscape globally. They cover approximately 43% of the earth's surface. There are different types of grasslands: meadow, steppe, and savannah.

In Morocco, grasslands are steppe-type. They are constituted, in general, from alfalfa and aromatic shrubs. They cover about 3,255,714 ha, which constitutes about a third of the national forest cover [1].

Rosemary, *Rosmarinus officinalis* L. [family LAMIA-CEAE], is an evergreen shrub native to the Mediterranean region. It belongs to the mint family. This resource has economic and biological importance. Hence, estimating the biomass is required and has many utilities, e.g., monitoring the biomass fluctuation to oversee the ecosystem changes, predicting the production for agricultural cooperatives operating in this field, and for the high commission of water and forest to fairly tax the incomes of those cooperatives.

A good estimation of the biomass starts with good mapping. In general, detecting vegetation cover in arid and semi-arid areas using remote sensing techniques is problematic. When the density of leaves is weak, the adjusted soil influences the spectral emittance and makes it difficult to detect the vegetation cover. Previous works propose the use of time-series data to differentiate between evergreen and deciduous species or to describe the seasonal change in agricultural lands [2]. Other works resort to the use of images captured in dry seasons [3].

Remote sensing techniques seem to be the practical way to detect and map vegetation canopy. They are largely used for this task. Those techniques perform simply, but effectively, by registering reflected electromagnetic waves [4]. Researchers exploit the vegetation spectral characteristic to discriminate it, directly from the original bands or by developing spectral indices.

Spectral indices (SIs) in remote sensing are mathematical equations that have as parameters wavelength bands. Each index is developed considering the reflectance characteristics of the objects. The most common form is the normalized difference.

To realize the potential of SIs, one can refer to the bibliography to find several important applications, e.g., monitoring land use/landcover change [5, 6], inspecting the water and air quality [7], in geology [8], and vegetation cover monitoring [9].

H. Chafik (✉) · M. Berrada
Department of Mathematics and Informatics, National High School of Arts and Crafts, University of Moulay Ismail, Meknes, Morocco
e-mail: chafik.hassan@edu.umi.ac.ma

A. Legdou · A. Amine
Department of Informatics, Logistics, and Mathematics, National High School of Applied Sciences, University Ibn Tofail, Kenitra, Morocco

S. Lahssini
National Forestry School of Engineers, Sale, Morocco

© The Author(s), under exclusive license to Springer Nature Switzerland AG 2022
F. Barramou et al. (eds.), *Geospatial Intelligence*, Advances in Science, Technology & Innovation,
https://doi.org/10.1007/978-3-030-80458-9_13

Remote sensing technics, including SIs, could be fully exploited using the machine learning (ML) approach. ML, as a part of artificial intelligence, designs a category of algorithm that enables machines to learn from data and become accurate in prediction and classification tasks. Nowadays, ML finds various applications, e.g., in image processing, data analysis, public health, and modeling phenomena.

Recently, several published works handled mapping and biomass estimation. By browsing the bibliography, we find a massive use of remote sensing data, either optical, lidar, and radar [10–12]. The authors used different algorithms in the mapping task, such as random forest, maximum likelihood, and spectral angle mapper [13–15]. They used different algorithms in the biomass estimation task either.

In this chapter, we aim to present a simple and efficient methodology to estimate the rosemary dry biomass. Concerning the mapping of the rosemary cover, we will exploit a previously published work. Then, we will demonstrate the biomass estimation method.

This paper presents an application on rosemary lands in the Errachidia region, southeast of Morocco. The methodology is scalable in larger areas and is applicable to other zones having similar conditions. The methodology is simple for reproduction by a large community, either experts, scientists, or end-users.

2 Materials and Methods

2.1 Sentinel 2 MSI Images

Sentinel 2 MSI satellite images are available since June 2015. They are captured using the multi-spectral instrument (MSI) carried on the two satellites 2A and 2B. The images contain 13 spectral bands at 10–60 m spatial resolution.

What makes Sentinel 2 MSI images useful is that they are orthorectified tiles covering 100 km² each. They provide 13 spectral bands sensitive to different earth surface objects which increases the information that could be extracted and the possibility to formulate more spectral indices in comparison to other free access data like the Landsat series.

Besides, more open-access software provide tools to treat those images such as QGis and ESA's SNAP Toolbox. These products are available via the Copernicus open access hub and USGS Earth explorer platform.

The problem that occurs in this stage is the date of the capture. In our region, precipitation is irregular. Hence, we cannot use images captured in summer as scientific papers suggest for better discrimination between evergreen and drought semi-deciduous vegetation [3] This irregularity could be seen by observing the last two years (2018–2019) precipitation records (Fig. 1). Several periods of drought and precipitation alternate during the year but not fitting with the seasons. Hence, summer does not necessarily define a dry season in this region.

The precipitation records show four dry periods. The longest one is P3. It reigned for five months (from October 2018 to February 2019). A second one (P1) was recorded from February 2018 to April 2018. It lasted for three months, while two short ones, P2 and P4, were registered in June and July 2018 and 2019, respectively.

Table 1 shows the available Sentinel 2 images during the last two months of P3. Twelve scenes are spanning 59 days with 6 days temporal resolution. The best scene should be as far as possible from the last precipitation and have less cloud cover percentage.

Table 2 shows further scenes from the last precipitation and having less cloud cover. Based on what was mentioned before, the S9 seems to be the best scene. It comes after 164 days of drought and has a weak cloud cover.

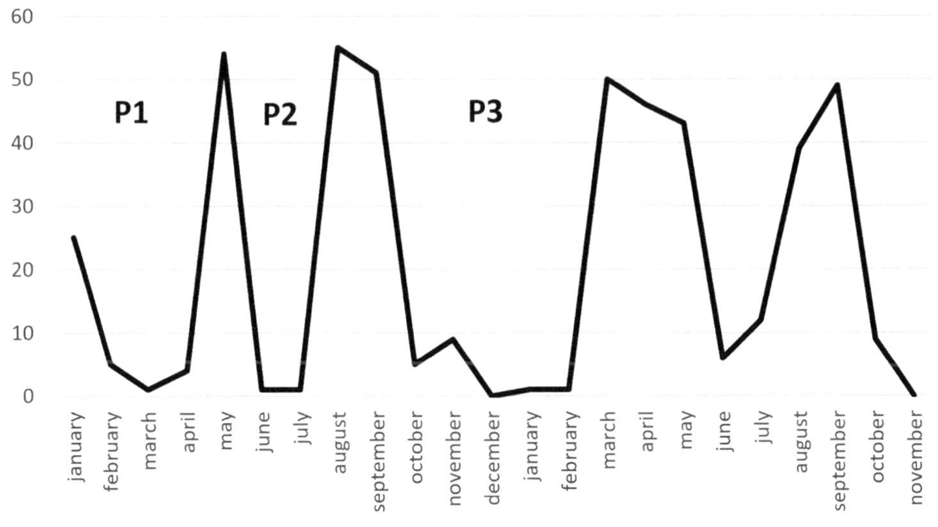

Fig. 1 Precipitation records of the study area during 2018–2019 (Reproduced from [16])

Table 1 The available Sentinel 2 MSI scenes during P3

Scene	S1	S2	S3	S4	S5	S6	S7	S8	S9	S10	S11	S12
Day	04	09	14	19	24	29	03	08	13	18	23	28
Month	January						February					
Year	2019											

Table 2 Sentinel 2 MSI scenes with the lowest cloud cover

Scene	S5	S8	S9
Date	2019/01/24	2019/02/08	2019/02/13
Cloud cover (%)	0.24	0.029	0.67

Table 3 Sentinel 2 MSI scenes with the lowest cloud cover

Formula	Author
$NDWI = \frac{NIR - SWIR}{NIR + SWIR}$	Hardisky et al. [19]
$NDWI = \frac{GREEN - NIR}{GREEN + NIR}$	McFeeters [20]

The image was clipped for the study area, radiometrically, and atmospherically corrected using the DOS tool in QGIS software.

2.2 Spectral Indices

The first one is the NDVI (normalized differenced vegetation index). It is a famous index based on detecting the gap between the red wave and the near-infrared wave reflectance for a given body. It is a biophysical parameter that correlates with chlorophyll concentration (photosynthetic activity) [17], so NDVI can give information on vegetation cover, and land cover in general, in an indirect way [18]. Its formula is (1):

$$NDVI = \frac{NIR - RED}{NIR + RED}. \qquad (1)$$

NIR is the near-infrared wave band, and RED is the red waveband.

NDWI (normalized differenced water index) is widely used for mapping water surfaces (lacs, water dam, etc.), but it is also used for describing the wetness of objects. This index was proposed by several authors with different formulas from which we cite those shown in Table 1 (Table 3).

NIR is the near-infrared waveband, SWIR is the short infrared waveband, and GREEN is the green waveband. We will adopt, in our paper, the McFeeters formula.

The third index is SAVI (Soil Adjusted Vegetation Index). It is an index proposed by [21], as a new spectral vegetation index. Its formula is (2):

$$SAVI = 1.5 + \frac{NIR - RED}{NIR + RED + 0.5} \qquad (2)$$

Those indices have enhanced the object's discrimination in the arid climate context. They will be considered as

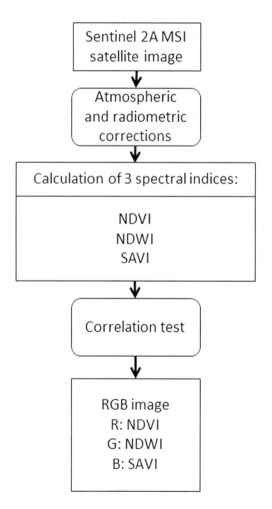

Fig. 2 Organigram presenting the indices calculation steps (Reproduced from [16])

Fig. 3 Organigram presenting the RF model establishment (Reproduced from [16])

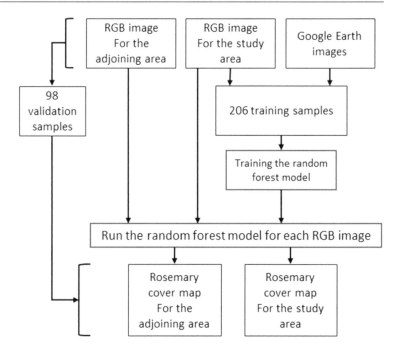

parameters for the RF model used to classify the rosemary cover. They will be stacked to form an RGB image. The methodology is resumed in Fig. 3 (Fig. 2).

2.3 Random Forest Model

Random decision forest was created by [22] and proposed by [23] as an ensemble of decision trees, where each tree is a classifier. It consists of a bagging procedure to generate random vectors from the original training set. Each vector feeds a classifier. The final decision to assign an element to a class is obtained by a majority vote among the trees. The combination of multiple decision trees in the RF algorithm reduces the overfitting [24]. As with any supervised classifier, training is required to build an efficient model.

The dataset contains 304 terrain truth samples. To each sample, three values relative to NDVI, NDWI, and SAVI indices were associated. The values are extracted from the RGB image in a GIS environment. Those indices are considered as the independent variables. Besides, a label (the dependent variable) is assigned to define the class to which belongs each sample.

To structure the RF model, three parameters are required to set: the number of trees, the homogeneity function, and the minimum number of splits. The number of trees is not evident to define. In general, the more trees we use, the better we get the results. But in some cases, the benefit of using an important number of trees could be lower than the

cost of time-consuming. Hence, it becomes impractical. We tested 50, 100, 200, and 300 trees and we got approximately the same training score and the same parameter's weights distribution (Table 4). It was decreasing as the number of trees increased.

Those results could be explained by the small training set we are dealing with. So, it is unnecessary to implement a large tree number. We kept 50 trees for our model.

A second detail to discuss is the function used for measuring the quality of a split. In our model, we used the Gini impurity function. It is a metric that measures, within the subsets, the homogeneity of the target variable. In other words, it measures the probability of misclassifying an object into a class [25]. Its formula is (3):

$$i_{Gini}(u) = \sum_{i=1}^{d} \frac{u_i}{\|u\|_1}\left(1 - \frac{u_i}{\|u\|_1}\right) \qquad (3)$$

The minimum number of samples required to split an internal node is two samples, so nodes are expanded until they are pure or until all leaves contain less than the minimum number of split samples.

After the training step, the model was run to classify the RGB image for the study area and distinguish four classes: very weak rosemary cover (VWR), weak rosemary cover (WR), important rosemary cover (IR), and bare soil (Bs). The model was run for an adjoining zone to test its scalability. For the validation step, we used 98 validation points taken from both areas.

Table 4 Training scores and weights distribution as a function of the number of trees ([16])

Trees number	50	100	200	300
Training score	0.728	0.723	0.699	0.708
Weights				
NDVI	0.414	0.408	0.423	0.424
NDWI	0.254	0.257	0.249	0.251
SAVI	0.331	0.333	0.326	0.324

2.4 Biomass Calculation

The last part of the work is to estimate the rosemary production by calculating the dry biomass. To reach this purpose, we went through the estimation of the green biomass, then the deduction of the dry one.

2.4.1 Tuff Weight

This step consists of estimating the average weight of a single rosemary tuff. The average weight could be simply the mean of the real weight samples or estimated using Eq. (4) that relies the weight to the diameter and the height of tuff samples, then calculate the mean estimated weight.

We went through the second proposition, and for this, 40 random tuff samples were withdrawn from the field. For each one, the diameter and height were measured. Rosemary tuffs do not present a perfect spheric form; therefore, two diameters were measured, then used to get the mean diameter. The diameters are not necessarily perpendicular.

$$B_e = 1.05(2.001\, D^{1.39551} H^{0.26943}) \qquad (4)$$

B_e is the estimated weight, D is the mean diameter of the tuff, and H is the height of the tuff.

Since we have 40 samples, the estimated weight will be:

$$Be_m = \sum_1^{40} \frac{B_e}{40} \qquad (5)$$

To test how accurate is our estimation, we compared Be_m with the real average weight Br_m using Eq. (6):

$$D = ((Br_m - Be_m)/Br_m) \times 100 \qquad (6)$$

2.4.2 Number of Tuffs Per Parcel

Now we need to get the thematic density classes (IR, WR, VWR), gotten previously, expressed in terms of the number of tuffs per unit area.

Within each class area, we toke 10 randomly distributed parcels sized of 0.010 ha (100 m^2) which is the same size as the image pixels. Then we calculated the number of tuffs within each parcel. Table 5 presents the detail of the calculation.

2.4.3 Wet and Dry Biomass Estimation

To get the biomass, we will follow a simple series of calculations. Within each class, for each parcel, the biomass $B_{p,class}$ will be the number of tuffs by the estimated average weight:

$$B_{p.class} = Be_m \times N_{mean} \qquad (7)$$

Then, the biomass of the class $B_{c,class}$ is $B_{p,class}$ by the total number of pixels $N_{pixels,class}$:

$$B_{c.class} = B_{p.class} \times N_{pixels.class} \qquad (8)$$

Hence, the total estimated biomass is the sum of $B_{c,class}$:

$$Bt_{wet} = \sum_{class=1}^{3} B_{c.class} \qquad (9)$$

We have now the green biomass estimation B_{wet}. We need to deduct the dry biomass. The efficient way is to define the dryness ratio D_r of the rosemary by comparing the wet weight B_e calculate before and the dry one B_{dry} measured after leaving the tuffs drying outdoor. This experience was elaborated for each tuff sample.

Table 5 Withdrawn parcels and the relative number of tuffs per parcel for each class

Class	Nbr of parcels	Min tuff/parcel (N_{min})	Max tuff/parcel (N_{max})	Mean tuff/parcel (N_{mean})
IR	10	16	22	19
WR	10	11	14	12.5
VWR	10	4	10	7

$$D_r = (B_{dry}/B_e) \qquad (10)$$

Then we calculate the mean dryness ratio Dr_m:

$$Dr_m = \sum_1^{40} \frac{D_r}{40} \qquad (11)$$

Finally, the dry biomass is:

$$Bt_{dry} = Bt_{wet} \times D_r \qquad (12)$$

2.5 Results

2.5.1 Spectral Indices

We start the exhibition of the results by describing the spectral indices maps. First, we have the NDVI map (Fig. 5). The values are ranging from −0.092 to 0.651. The lowest values correspond to bare soil (red color), the highest values refer to important vegetation cover, and intermediate values are reflecting different vegetation cover densities (Fig. 4).

The second map is the NDWI map (Fig. 6). The values are ranging from −0.644 to 0.074. The lowest values correspond to dry vegetation because it reflects the near-infrared waves largely more than the green waves, so the difference GREEN–NIR will be very low. As the vegetation cover gets weaker, the influence of the soil becomes important. As a result, the value of the NDWI will increase to get its maximum for the bare soil.

The last map is relative to the SAVI parameter (Fig. 7). For this index, we had values from 0.051 to 0.187. All the values are positive but close to zero. The highest values refer to bare soil, while the lowest ones refer to the existence of a vegetation cover.

Those indices showed different correlations (Table 6). The NDVI is mediumly negatively correlated to NDWI, while it is weakly positively correlated to SAVI. NDWI and SAVI showed a relatively important positive correlation. Those correlations are important to verify, to make sure that none of the variables are fully correlated, so all the variables chosen will contribute differently to explaining the phenomenon.

2.5.2 Random Forest Classifier

The RF model classified the RGB image and gave as results a density map showing four classes (IR, WK, VWK, and Bs), for the study area (Fig. 8), and the area used for validation (Fig. 9).

The validation of the classification results gave a score of 90%. The validation has so proven the sensibility of the model toward the weak rosemary cover.

By comparing the truth terrain samples and the results obtained, we observe that the classes showed a high matching rate (Table 7). The model was good in detecting weak rosemary cover class (WR), and in second place, in detecting important rosemary cover class (IR), then respectively for VWR and Bs class. We noticed that the VWR class was mingled with Bs class.

Fig. 4 Map of NDVI index calculated for the study area (Reproduced from [16])

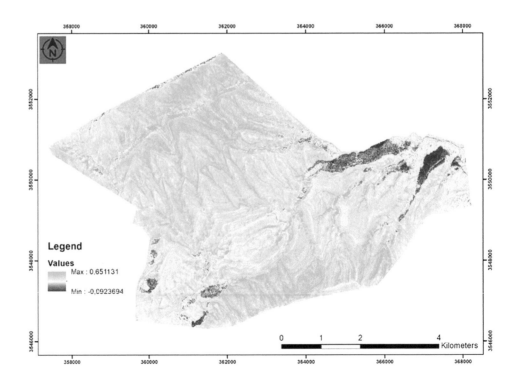

Fig. 5 Map of NDWI index calculated for the study area (Reproduced from [16])

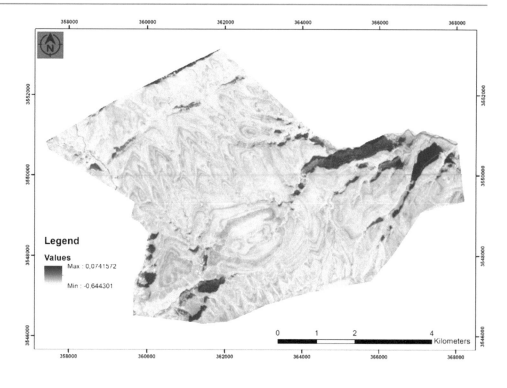

Fig. 6 Map of SAVI index calculated for the study area (Reproduced from [16])

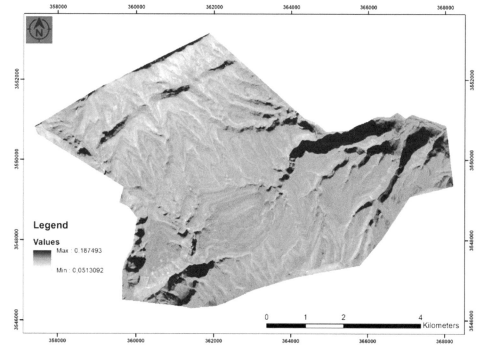

2.5.3 Dry Biomass Estimation

Respectively to the methodology, we calculated the average weight Be_m of the single tuff. Next, we calculated the wet biomass and finally, we calculated the dry biomass of the study area. The results are detailed in Tables 8 and 9.

2.6 Conclusion

The approach presented in this work, based on remote sensing techniques and machine learning, showed its utility in building a classification model capable to classify weak

Fig. 7 Map of rosemary cover established by the model for the study area (Reproduced from [16])

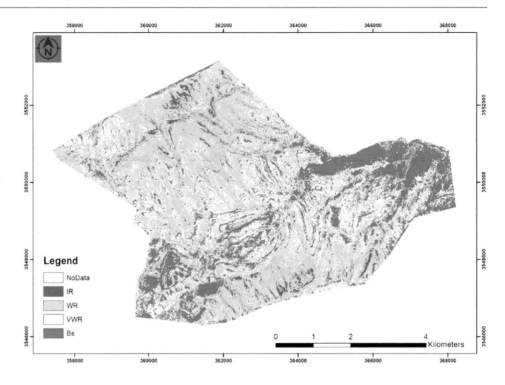

Table 6 Correlation between spectral indices (Chafik et al. [16])

	NDVI	NDWI	SAVI
NDVI	1	−0.52	0.36
NDWI		1	−0.79
SAVI			1

Fig. 8 Map of rosemary cover established by the model for the adjoining zone (Reproduced from [16])

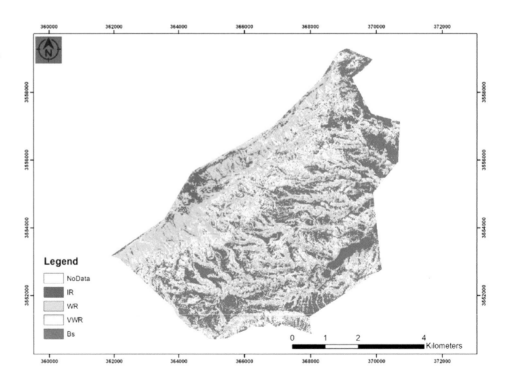

Table 7 Validation score relative to each class ([16])

Class	Match	Mis match	Match %
IR	19	3	86.36
WR	26	0	100
VWR	20	7	74.07
Bs	18	5	78.26

Table 8 Estimated and real average weights and relative accuracy

Be_m (kg)	Br (kg)	D (%)
0.05	0.053	7

Table 9 Calculation detail of the density, the estimated wet and dry biomass for each class with the total

Class	Area (ha)	Density (tuff / ha)	Bt_{wet} (tons)	Bt_{dry} (tons)
IR	218.4	1900	414.96	248.97
WR	1488.26	1250	1860.325	1116.19
VWR	1316.74	700	921.718	553.03
Total	3023.4	–	**3188.003**	**1918.19**

vegetation cover. The remote sensing technics provided free and easily manipulated parameters for the model. The random forest classifier algorithm has proven its potential in distinguishing rosemary different densities, based on only three spectral indices. The good quality of the map issued from the classification gave accurate support for the biomass estimation. All those qualities make the present methodology practical and easy for reproduction by users, either researchers or any interested user.

Acknowledgments The authors of this work wish to express big recognition to the Regional Direction of Water and Forests and Luting against Desertification of Middle Atlas-Meknes, Morocco for field accompaniment and data providing.This work is supported by the University of Moulay Ismail, Meknes, Morocco, and the National Center for Scientific and Technical Research (CNRST).

References

1. P. Blérot, O. Mhirit, *Le grand livre de la forêt marocaine*. Mardaga (1999)
2. Y. Yang, T. Wu, S. Wang et al., The NDVI-CV Method for mapping evergreen trees in complex urban areas using reconstructed landsat 8 time-series data. Forests **10**, 1–16 (2019). https://doi.org/10.3390/f10020139
3. F. Maselli, A. Rodolfi, S. Romanelli et al., Classification of Mediterranean vegetation by TM and ancillary data for the evaluation of fire risk. Int. J. Remote Sens. **21**, 3303–3313 (2000). https://doi.org/10.1080/014311600750019912
4. J. Xue, B. Su, Significant remote sensing vegetation indices: a review of developments and applications. J. Sens. **2017**(2017). https://doi.org/10.1155/2017/1353691
5. M. Mathan, M. Krishnaveni, Monitoring spatio-temporal dynamics of urban and peri-urban land transitions using ensemble of remote sensing spectral indices-a case study of Chennai metropolitan area, India. Environ. Monit. Assess. **192**, 15 (2019). https://doi.org/10.1007/s10661-019-7986-y
6. C. Polykretis, M.G. Grillakis, D.D. Alexakis, Exploring the impact of various spectral indices on land cover change detection using change vector analysis: a case study of Crete Island. Greece. Remote Sens. **12**(2020). https://doi.org/10.3390/rs12020319
7. V. Sagan, K.T. Peterson, M. Maimaitijiang et al., Monitoring inland water quality using remote sensing: potential and limitations of spectral indices, bio-optical simulations, machine learning, and cloud computing. Earth-Sci. Rev. **205**(2020). https://doi.org/10.1016/j.earscirev.2020.103187
8. A. Khalifa, Z. Çakır, Ş Kaya, S. Gabr, ASTER spectral band ratios for lithological mapping: a case study for measuring geological offset along the Erkenek Segment of the East Anatolian Fault Zone. Turkey. Arab. J. Geosci. **13**, 832 (2020). https://doi.org/10.1007/s12517-020-05849-y
9. A.M. Akhtar, W.A. Qazi, S.R. Ahmad et al., Integration of high-resolution optical and SAR satellite remote sensing datasets for aboveground biomass estimation in subtropical pine forest. Pakistan. Environ. Monit. Assess. **192**, 584 (2020). https://doi.org/10.1007/s10661-020-08546-1
10. Y. Li, M. Li, C. Li, Z. Liu, Forest aboveground biomass estimation using Landsat 8 and Sentinel-1A data with machine learning algorithms. Sci. Rep. **10**, 1–12 (2020). https://doi.org/10.1038/s41598-020-67024-3
11. D. Deb, S. Deb, D. Chakraborty et al., Aboveground biomass estimation of an agro-pastoral ecology in semi-arid Bundelkhand region of India from Landsat data: a comparison of support vector machine and traditional regression models. Geocarto Int. **1–16** (2020). https://doi.org/10.1080/10106049.2020.1756461
12. S. Yadav, H. Padalia, S.K. Sinha et al. Above-ground biomass estimation of Indian tropical forests using X band Pol-InSAR and Random Forest. Remote Sens. Appl. Soc. Environ. **21**, 100462 (2021). https://doi.org/10.1016/j.rsase.2020.100462

13. S. Jarradi, K. Tounsi, The use of satellite remote sensing and geographic information systems in monitoring the dynamics of alfatières aquifers . Case of the delegation of Hassi el Frid of the governorate of Kasserine in L'utilisation de la télédétection satellitaire et des s. CI:3449–3458 (2018)

14. G.S. Adjognon, A. Rivera-Ballesteros, D. van Soest, Satellite-based tree cover mapping for forest conservation in the drylands of Sub Saharan Africa (SSA): application to Burkina Faso gazetted forests. Dev. Eng. **4** (2019). https://doi.org/10.1016/j.deveng.2018.100039

15. U.N.T. Nguyen, L.T.H. Pham, T.D. Dang, An automatic water detection approach using Landsat 8 OLI and Google Earth Engine cloud computing to map lakes and reservoirs in New Zealand. Environ. Monit. Assess. **191** (2019). https://doi.org/10.1007/s10661-019-7355-x

16. H. Chafik, M. Berrada, A. Legdou, Exploitation of spectral indices NDVI, NDWI & SAVI in Random Forest classifier model for mapping weak rosemary cover: application on et al., IEEE International Conference on Moroccan Geomatics. MORGEO (2020). https://doi.org/10.1109/Morgeo49228.2020.9121895

17. P.J. Sellers, Canopy reflectance, photosynthesis and transpiration. Int. J. Remote Sens. **6**, 1335–1372 (1985). https://doi.org/10.1080/01431168508948283

18. Q. Wang, J.D. Tenhunen, Vegetation mapping with multitemporal NDVI in North Eastern China Transect (NECT). Int. J. Appl. Earth Obs. Geoinf. **6**, 17–31 (2004). https://doi.org/10.1016/j.jag.2004.07.002

19. M.A. Hardisky, V. Klemas, R.M. Smart, The influence of soil salinity, growth form, and leaf moisture on the spectral radiance of Spartina alterniflora canopies. Photogramm. Eng. Remote Sens. **49**, 77–83 (1983)

20. S.K. McFeeters, Using the normalized difference water index (ndwi) within a geographic information system to detect swimming pools for mosquito abatement: a practical approach. Remote Sens. **5**, 3544–3561 (2013). https://doi.org/10.3390/rs5073544

21. A.R. Huete, A soil-adjusted vegetation index (SAVI). Remote Sens. Environ. **25**, 295–309 (1988). https://doi.org/10.1016/0034-4257(88)90106-X

22. T.K. Ho, Random decision forests. Proc. Int. Conf. Doc. Anal. Recognit. ICDAR **1**, 278–282 (1995). https://doi.org/10.1109/ICDAR.1995.598994

23. L. Breiman, Random forests-random features (1999)

24. A. Subudhi, M. Dash, S. Sabut, Automated segmentation and classification of brain stroke using expectation-maximization and random forest classifier. Biocybern. Biomed. Eng. **40**, 277–289 (2020). https://doi.org/10.1016/j.bbe.2019.04.004

25. E. Laber, L. Murtinho, Minimization of Gini impurity: NP-completeness and approximation algorithm via connections with the k-means problem. Electron. Notes Theor. Comput. Sci. **346**, 567–576 (2019). https://doi.org/10.1016/j.entcs.2019.08.050

Printed in the United States
by Baker & Taylor Publisher Services